Causes and Risks for Autism

Causes and Risks for Autism

Alessia C. Giordano
and
Viola A. Lombardi
Editors

Nova Biomedical Books
New York

Library of Congress Cataloging-in-Publication Data

Causes and risks for autism / [edited by] Alessia C. Giordano and Viola A. Lombardi.
p. ; cm.
Includes bibliographical references and index.
ISBN 978-1-60456-861-5 (hardcover)
1. Autism. I. Giordano, Alessia C. II. Lombardi, Viola A.
[DNLM: 1. Autistic Disorder--etiology. 2. Risk Factors. WM 203.5 C374 2009]
RC553.A88C38 2009
616.85'882--dc22

2008049484

Published by Nova Science Publishers, Inc. ✝ New York

Contents

Preface

The terms 'autism' and 'autistic' derive from the Greek word *autos*, meaning self. This is appropriate to describing the autistic behavioral phenotype in which there is a pathological impairment in socialization and verbal and nonverbal communication, in addition to behavior and interests that are often highly restricted and repetitive. The autistic individual often appears isolated, and unable to make sense of the world around them. They often reveal an inability to predict and understand the behavior of others, and perceptions of the world remain fragmented and are not embedded into a coherent pattern or structure. This book discusses the causes and risks of autism from researchers around the globe.

Chapter I - The central and peripheral serotonin (5-hydroxytryptamine, 5-HT) systems are discussed in a single conceptual framework. It is hypothesized that the abnormal development of the autistic brain and the blood hyperserotonemia of autism are caused by factors that are expressed in both 5-HT systems. The underlying biology of these processes is explored in mathematical models of blood 5-HT levels. Predictions of the models are discussed in relation to the early, unobservable development of the autistic brain.

Chapter II - Genetic heterogeneity for a multifactorial disease such as autism would imply that any two patients are unlikely to share the same susceptibility loci. Since different susceptibility loci may affect the same function and since functions may be considered at different levels of analysis, the following question arises: at what phenotypic levels is convergence attained by different genes in autism ? This is an important question to answer in order to shed light on subcellular, cellular or multicellular structures and functions possibly involved in the pathogenesis of autism. Among the various attempts made to answer this question it is worth mentioning the use of the best available genetic information to focus on the formation, structure and function of the glutamatergic synapse. How could we use this working scheme in a more efficient and hopefully fruitful way ?

The recent advent of genomic technology offers an unprecedented opportunity to study the biological structure of autism deep into its molecular roots. For example, recent work on structural genomics has led to the identification of many Copy Number Variations in the gene-rich regions of the genome detected in a large number of patients. Interesting advances have also been made by analyzing the global gene-expression profiles in blood cells of autistic patients. The large amount of data generated by these studies is likely to contain precious information about the pathogenesis of autism. In this chapter, the authors we review the literature on the genomic studies in ASD and present the results of a preliminary study showing how candidate biological processes for ASD can be identified from microarray data using the Gene Ontology database. The authors also discuss the importance of the transition from a strategy based on the search for candidate genes to the search for candidate ontology terms to expand our current understanding on the biological processes impaired in ASD.

Chapter III - Gastropods have an unusual genetic mechanism in which the mother's genotype determines the shell phenotype of her offspring. Experiments have shown that these unknown, maternal-effect genes determine shell chirality, the direction each snail shell coils, with dextral or right handed having a clockwise spiral, and sinistral or left handed having a counterclockwise spiral. Most snail species consist only of dextral individuals, while some species are entirely sinistral, and some species are polymorphic for shell chirality, producing both dextrals and sinistrals in the same species. This particular genetic mechanism implies a lack of homozygous-dominant (*R/R*) individuals in the alternative shell phenotype, while the primary phenotype can have any of the genotypes of homozygous dominant, heterozygous (*R/r*), or homozygous recessive (*r/r*) in snails. In humans, the hair-whorl rotation, the direction in which the hair spins at the back of the head, is generally either clockwise or counterclockwise and has been associated with handedness, cerebral laterality, and sexual orientation. As such, direction of hair-whorl rotation is thought to be a phenotypic marker underlying the genetics of various behavioral phenotypes in our species. Previous research has already associated handedness with sexual orientation, psychosis, and autism spectrum disorders (ASD), but to the best of my knowledge, hair-whorl rotation has not been tabulated in autistic individuals. A recent theory has proposed that maternal-effect genes are involved in determining handedness and hair-whorl rotation in humans, in addition to various behavioral phenotypes in our species. These genes (possibly *RHD* and *RHCE*) predict a lack of homozygous-dominant individuals among the alternative phenotypes in humans, analogous to shell chirality in snails. If maternal-effect genes are interacting with biparentally-expressed genes to cause some cases of ASD and other behavioral phenotypes in humans, then current genomic search results (i.e., linkage mapping using single nucleotide polymorphisms) for all of these genes will be obscured. To increase precision of genome searches during sibling studies, researchers should search for maternal-effect genes (i.e., never or almost never homozygous in the alternative phenotype; alleles shared *less* commonly than expected by chance in sets of affected siblings) in addition to biparentally-expressed genes (i.e., genes always or almost always homozygous in the alternative phenotype; alleles shared *more* commonly than expected by chance in siblings). If it can be shown that the snail chirality gene is homologous to the human handedness and hair-whorl genes, even if these genes turn out not to be *RH* genes, snails may still provide an animal model for evaluating possible environmental causes of ASD in humans. For example, aquatic snails, many of which have a short generation time and breed easily in captivity, could be treated with mercury to test the expression of the opposite chirality in their offspring. Other chemical agents or mixtures could also be tested in snails to determine possible correlates to environmental exposure in humans.

Chapter IV - The terms 'autism' and 'autistic' derive from the Greek word *autos* meaning self. This is appropriate to describing the autistic behavioral phenotype in which there is a pathological impairment in socialization and verbal and nonverbal communication, in addition to behavior and interests that are often highly restricted and repetitive (the triad; American Psychiatric Association, 1994). The autistic individual often appears isolated, and unable to make sense of the world around them. They often reveal an inability to predict and understand the behavior of others, and perceptions of the world remain fragmented and are not embedded into a coherent pattern or structure. Time is part of the fundamental intellectual structure in which we make sense of the events in our lives. 'Timing and time perception allow us to unite action sequences and events occurring separately in time, to adapt to reoccurring situations, and to predicate behavior on what is expected to occur'. Timing and time perception are essential for adaptation and learning, memory and attention, cognitive development, and social synchrony and communication (see Meck, 2003). Firsthand accounts of people with autism often report a need to adhere to rituals or routines to compensate for a failure to predict events, and to their disorientation in time. They reveal a general lack of understanding about the passage of time, and appear stuck in the present.

It is for these reasons that the issue of timing and time perception in autism is particularly intriguing. The authors will review empirical evidence that collectively suggests time perception may be disordered in autism, and postulate that fundamentally, a disturbed 'time sense' may contribute to features of the autistic behavioral phenotype.

Chapter V - Language abnormalities are critical in diagnosing autism, including the absence of or severe language delay, inability to initiate or sustain a conversation and the use of stereotypic or repetitive language. Symptoms related to language may vary in intensity; however, the common denominator (i.e., an inability to establish effective communication) is ever present in these patients.

Other diagnostic criteria address deficiency in or lack of social interaction, reflected in abnormal play behavior or in the manipulation of elements having inappropriate, repetitive patterns of a non-functional ritualistic nature.

Children suffering from autism generally present no major alterations of early motor development during the first and second years of life and the child is able to sit, crawl and walk within the expected time-frames. Retrospective studies of films of autistic children have shown subtle differences in object manipulation, visual attention to social stimuli as well as in smiling and exploration. However, the alarm signal (and the one usually triggering consultation) is the absence of or defective language development.

The above observations indicate that the central alteration in an autistic child resides in the ability to relate to others, particularly involving verbal and non-verbal communication, difficulties in establishing social contact and an inability to detect the intentions of others.

Neuro-anatomical studies and functional magnetic resonance imaging have shown structural variations and alterations involving the cerebellum, as well as temporal and frontal lobes. One of the striking features in these patients from the clinical viewpoint is the absence of alerting reaction and response to the human voice, while the orientating reflex to environmental sound stimuli remains normal. Dysphasia may range from lack of recognizing basic units or phonemes to problems in integrating a particular word with its meaning. The clinical picture becomes more complex as a child grows older and more participation in activities requiring social interaction is expected.

Early recognition and appropriate treatment of difficulties with processing social information provides a basis for treatment methodology that is often able to change these patients' worsening clinical course.

Short Communication - Histidinemia is one of the most frequent errors of amino acid metabolism. Metabolic blockage of histidase activity increases histidine concentrations in body fluids. There have been a lot of controversies whether histidinemia is a harmless biochemical error or a neurophycological disease. Several investigations demonstrated that histidinemic patients often presented autistic features, but others showed that they were asymptomatic. The authors investigate on 70 patients with histidinemia whether they are accompanied with autistic symptoms or not. Ten patients (14.3% of 70 patients) were diagnosed as having pervasive developmental disorder (DSM-IV). In detailed classification, five patients (7.1%) were autistic disorder, four (5.7%) were Asperger's disorder, one (1.4%) was pervasive developmental disorder not otherwise specified. The present study confirmed the frequency of pervasive developmental disorder was extremely high in histidinemic patients. The authors hypothesize that patients should present autistic symptoms if histidinemia and other factors (whichever genetic or environmental factor) are combined; that is histidinemia may be a risk factors for autism spectrum disorders.

In: Causes and Risks for Autism
Editors: A. C. Giordano et al.
ISBN: 978-1-60456-861-5

Chapter I

Autism as a Theoretical Problem and the Significance of Blood Hyperserotonemia

Skirmantas Janušonis*
Department of Psychology, University of California,
Santa Barbara, CA 93106-9660

1. Abstract

The central and peripheral serotonin (5-hydroxytryptamine, 5-HT) systems are discussed in a single conceptual framework. It is hypothesized that the abnormal development of the autistic brain and the blood hyperserotonemia of autism are caused by factors that are expressed in both 5-HT systems. The underlying biology of these processes is explored in mathematical models of blood 5-HT levels. Predictions of the models are discussed in relation to the early, unobservable development of the autistic brain.

* Skirmantas Janušonis: Tel: 805-893-6032; FAX: 805-893-4303; E-mail: janusonis@psych.ucsb.edu

2. Abbreviations

5-HT, 5-hydroxytryptamine (serotonin);
5-HTP, 5-hydroxytryptophan;
5-HIAA, 5-hydroxyindoleacetic acid;
BBB, blood-brain barrier;
CSF, cerebrospinal fluid;
EC cells, enterochromaffin cells;
ENS, enteric nervous system;
SERT, serotonin transporter (also abbreviated 5-HTT).

3. Introduction

The complexity of autism spectrum disorders (ASDs) arises from two very different sources. First, there is a distinct possibility that we are dealing with a number of biologically different disorders, even if only "classic" autism is considered. While the *Diagnostic and Statistical Manual of Mental Disorders* (DSM-IV-TR) remains the best of the bad options, currently we have no real evidence that we are not trying to understand the causes of the general limping disorder. The extent of the genetic complexity of ASDs has been demonstrated by a number of recent studies (Geschwind, 2007; Sebat et al., 2007; Szatmari et al., 2007; Abrahams and Geschwind, 2008; Lintas and Persico, 2008). While genetic complexity does not necessarily mean structural or functional complexity (Babloyantz, 1986; Nicolis and Prigogine, 1989; Geschwind, 2008), ASDs may well represent a number of biologically different conditions. The second source of the difficulty is our haphazard and inexcusably fragmented approaches, as insightfully observed by Belmonte et al. (2004). This situation has some objective causes: in autism, one has to step out the central nervous system (CNS) to see the entire picture. This comes at a price, as it involves understanding such seemingly unrelated and vast systems as the gut and the blood. Conversely, researchers specializing in the latter systems may be reluctant to add the CNS to the complexities of their own field.

The main purpose of this chapter is to present a conceptual approach I have been developing in the past several years. At its core, this approach is an attempt to bring several subfields of autism research into a single conceptual framework. I will also briefly review some basic facts about the "central" and "peripheral" serotonin systems, with the aim of furthering collaboration between researchers working on only one of these systems.

4. The Unobservable Autistic Brain

To date, a remarkably wide range of abnormalities has been reported in autistic brains, such as reduced numbers of Purkinje cells in the cerebellum (Williams et al., 1980; Ritvo et al., 1986; Kern, 2003), abnormal cortical minicolumns (Casanova et al., 2002; Casanova et al., 2003; Casanova et al., 2006; Casanova, 2007), abnormalities of the limbic system (Raymond et al., 1996; Amaral et al., 2003; Schumann et al., 2004), dysfunction of mirror neurons (Dapretto et al., 2006), alterations of cranial nerve nuclei (Rodier et al., 1996; Rodier, 2002), abnormal patterns of brain growth (Courchesne et al., 2004; Courchesne et al., 2005; Courchesne et al., 2007), inflammation of the brain (Vargas et al., 2005), and others (Kemper and Bauman, 2002; Palmen et al., 2004; Pickett and London, 2005; DiCicco-Bloom et al., 2006; Amaral et al., 2008). Due to the expected difficulty obtaining well-preserved autistic brain specimens, some of these reported alterations still await independent confirmation by other groups of researchers. Nevertheless, the evidence suggests that many neuroanatomical regions are affected in autistic brains.

By the time an autistic brain is available for examination in MRI or *postmortem* studies, the individual is very likely to have already been diagnosed with autism. This currently means that the individual is at least two years of age. Because of this late diagnosis, early (prenatal and perinatal) brain abnormalities are likely to escape direct detection in clinical and experimental studies. Importantly, these early abnormalities may be well-defined and few, but they may cause secondary alterations in many brain areas as the brain develops. Such early abnormalities would be virtually inaccessible to direct observation (because such fetal and perinatal brains would not yet be known to be autistic).

Given this situation, a successful approach may require (i) focusing on a consistent biological finding in diagnosed autistic individuals (phenomenon *B*), (ii) tracing its cellular and molecular origin (phenomenon *A*), and (iii) exploring whether phenomenon *A* may also affect early brain development (figure 1). To date, the most consistent biological alteration has been found in the blood of autistic individuals. Specifically, a large number of studies have shown that ethnically diverse autistic groups have elevated mean serotonin (5-

hydroxytryptamine, 5-HT) levels in blood platelets (Schain and Freedman, 1961; Hanley et al., 1977; Anderson et al., 1987a; Anderson et al., 1990; Cook, 1996; McBride et al., 1998; Anderson et al., 2002b; Coutinho et al., 2004; Mulder et al., 2004; Coutinho et al., 2007; Hranilovic et al., 2007; Hranilovic et al., 2008; Melke et al., 2008). While the biological causes of autism remain unknown, the platelet hyperserotonemia of autism currently is one of the most well-replicated findings in all of biological psychiatry (Anderson, 2002).

In contrast to the brain, blood platelets survive only a few days and are constantly replaced with new platelets. This means that the process that overloads blood platelets with 5-HT continues to be active years after the brain has developed. Some of these same molecular factors may play key roles in the development of the autistic brain, well before we can observe the brain in its final altered form. Therefore, blood platelets may serve as a window into the unobservable past of the autistic brain (figure 1). In this respect, the platelet hyperserotonemia of autism may be a biological analogy of the cosmic microwave background in astrophysics.

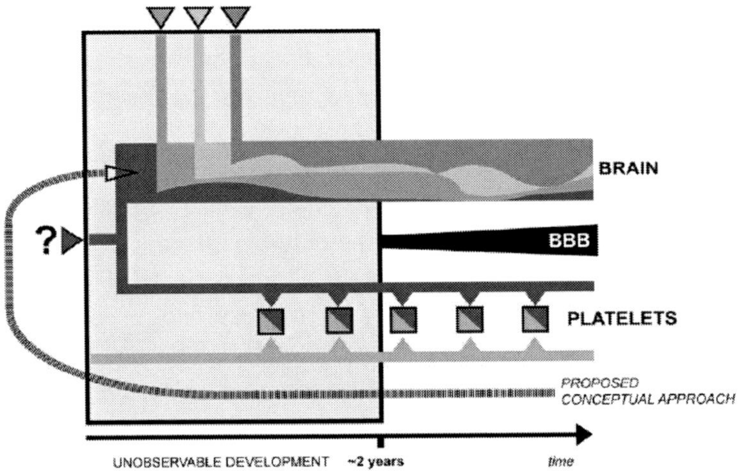

Figure 1. The platelet hyperserotonemia of autism may be caused, in part, by a factor that is also expressed in the developing autistic brain. The central and peripheral 5-HT systems are separated by the blood-brain barrier (BBB) that matures after birth. It is usually not clear until two years of age whether the brain is autistic. By that time, the alteration that set the developing brain on the autistic trajectory may no longer be evident. The factor that caused this alteration may continue to operate years after birth in the peripheral 5-HT system, where with other factors it may determine 5-HT levels in short-lived blood platelets (Janušonis, 2005; modified).

Before I present this general idea in more rigorous terms, a brief review of the mammalian serotonin systems is useful.

5. Central and Peripheral 5-HT Systems

Serotonin, its receptors and the serotonin transporter (SERT) play important roles in the embryogenesis of vertebrate and invertebrate species, before the CNS develops (Buznikov, 1984; Buznikov et al., 2001; Buznikov et al., 2005; Levin et al., 2006). Later in development, two 5-HT systems emerge in mammals (figure 2). We shall call them the *central* and *peripheral* 5-HT systems. It should be noted that these names are CNS-centric and may be misleading. These two 5-HT systems may also be named the CNS and non-CNS 5-HT systems, or 5-HT systems II and I, respectively.

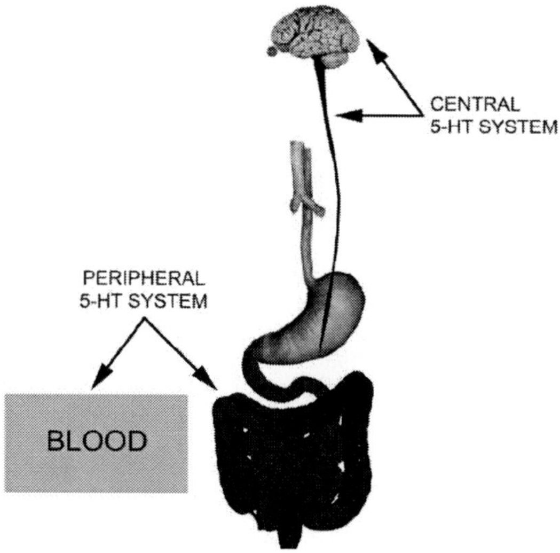

Figure 2. The central and peripheral 5-HT systems. Images from Nolte (2002) and Rohen et al. (2006) were used in the composite figure.

In both systems, 5-HT is synthesized from L-tryptophan, an amino acid. In both systems, most 5-HT is removed by converting it to 5-hydroxyindoleacetic acid (5-HIAA). Tryptophan can cross the blood-brain barrier (BBB), but its entry into the brain is limited by the competition among a few neutral amino acids (tryptophan is one of them). In contrast, 5-HT cannot cross the BBB, which means there are two separate 5-HT pools in the body: one in the CNS and the other outside the CNS. It is important to note that, strictly speaking, there is much peripheral 5-HT flowing through the brain in the immediate vicinity of neurons, but this 5-HT cannot escape blood capillaries. Blood capillaries are so tightly packed that in the cerebral cortex no neuron is more than some 100 μm away from a capillary (Duvernoy et al., 1981). Some brain serotonergic neurons can monitor carbon dioxide levels in brain blood vessels (Bradley et al., 2002; Severson et al., 2003).

The BBB is actually a number of barriers. These barriers can be defined based on the anatomical presence of tight junctions in blood capillaries, the functional permeability of blood capillaries for various substances, and other components that may or may not be present at the same time (Wenzel and Felgenhauer, 1976; Virgintino et al., 2004; Ge et al., 2005). In the human brain, tight junctions develop prenatally (Mollgard and Saunders, 1986; Bell et al., 1991; Virgintino et al., 2004).

Importantly, the BBB is a dynamic barrier. In this respect, the BBB resembles a set of immigration policies rather than a wall. The BBB is modulated by drugs, stress and other factors (Sharma, 2004a). The BBB is also altered by luminal and abluminal application of 5-HT to brain blood vessels; this process is mediated by 5-HT_2 receptors (Sharma, 2004b). Interestingly, the concentrations of free 5-HT on the inside and outside of brain capillaries may be comparable (Anderson et al., 1987b; Beck et al., 1993; Adell et al., 2002; Anderson, 2007), but without more accurate experimental measurements such comparisons should be treated with caution.

6. Central 5-HT System

In the central 5-HT system, 5-HT is synthesized by serotonergic neurons, most of which are located in the raphe nuclei of the brainstem. In this system, the enzyme that converts tryptophan into 5-hydroxytryptophan (5-HTP), the immediate precursor of 5-HT, is tryptophan hydroxylase 2 (*Tph2*) (Walther et al., 2003). In the central 5-HT system, 5-HT acts as a signaling molecule that can be detected by (perhaps all) neurons and by some glial cells, such as astrocytes (Whitaker-Azmitia, 2001). Serotonergic fibers (axons), originating in the raphe nuclei, spread throughout the brain, which becomes virtually embedded in a serotonergic meshwork. Even though it is often assumed that 5-HT is released from serotonergic fibers "diffusely", 5-HT signaling among neurons may occur through conventional synapses (Papadopoulos et al., 1987; Parnavelas and Papadopoulos, 1989). Extracellular 5-HT is pumped by SERT back into serotonergic fibers, where the 5-HT may be recycled. The extracellular 5-HT concentration in the brain is very low. In the rostral raphe nuclei, it has been estimated to be around 2 – 8 nM and may not be high enough to activate even 5-HT_{1A} autoreceptors unless 5-HT levels become excessive (Adell et al., 2002).

During brain development, some neurons may express tryptophan hydroxylase 1 (*Tph1*) (Nakamura et al., 2006). Also, developing thalamocortical neurons (and some other non-serotonergic neurons) can transiently express SERT and take up 5-HT, even though they themselves do not synthesize 5-HT (Lebrand et al., 1996; Lebrand et al., 1998; Lebrand et al., 2006). The exact pattern of the transient SERT expression shows species-specific variation (Lebrand et al., 2006).

The pineal gland uses *Tph1* to synthesize 5-HT (Walther and Bader, 2003), which can then be converted to melatonin. Melatonin can cross the BBB. However, in several respects the pineal gland lies outside the brain with regard to the BBB and, therefore, is a part of the peripheral 5-HT system.

In the central 5-HT system, 5-HT is converted to 5-HIAA that enters the cerebrospinal fluid (CSF) and that can be measured in lumbar puncture samples. The concentration of 5-HIAA in the human CSF has been estimated to be around 122 nM (Narayan et al., 1993). These 5-HIAA levels can be used to indirectly assess 5-HT function in the CNS (Anderson et al., 1988; Narayan et al., 1993). It has recently been suggested that CSF 5-HT levels, when analyzed with great care to minimize blood contamination, may provide a more direct and accurate measure of extracellular 5-HT in the CNS (Anderson et al., 2002a). The concentration of 5-HT in the CSF of *Macaca mulatta* has been estimated to be around 87 ng/L (Anderson et al., 2002a).

7. Peripheral 5-HT System

In the peripheral 5-HT system, most 5-HT is synthesized by enterochromaffin (EC) cells and neurons of the gut. Over 95% of the body's 5-HT is in the gut and over 90% of this 5-HT is stored in EC cells that are distributed in the enteric epithelium from the stomach through the colon (Gershon, 2004). The enzyme that converts tryptophan into 5-HTP in EC cells is tryptophan hydroxylase 1 (*Tph1*);

in gut neurons, it is tryptophan hydroxylase 2 (*Tph2*) (Walther et al., 2003; Gershon and Tack, 2007). Functionally, the 5-HT produced in the gut plays two distinct roles:

First, 5-HT produced in the gut serves as a signaling molecule (neurotransmitter) in the enteric nervous system (ENS) (Gershon, 2003) that contains as many neurons as the spinal cord (Gershon, 2004). Extracellular 5-HT is taken up by neuronal and non-neuronal cells of the gut that express SERT (Gershon, 2003; Gershon, 2004). A recent study has shown that SERT is important in intestinal immune and inflammatory responses; abnormal SERT function may increase the severity of immune-mediated colitis (Bischoff et al., 2009).

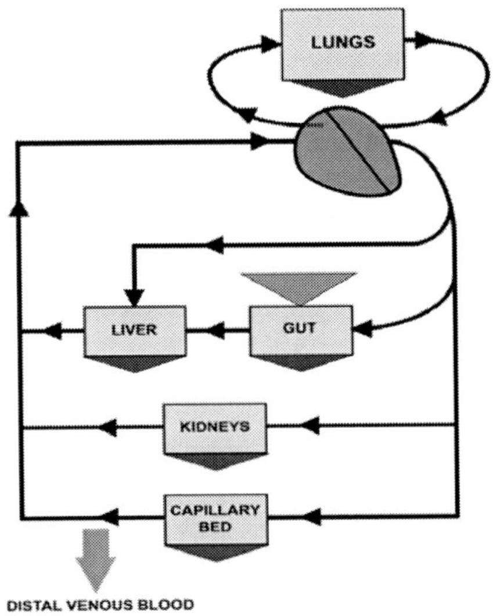

Figure 3. The circulation of free 5-HT in the peripheral 5-HT system. The triangle above the gut box represents gut 5-HT production, the triangles below boxes represent 5-HT clearance. In humans, blood 5-HT (free and sequestered in blood platelets) is usually measured in distal venous blood samples (wide arrow). Based on Anderson et al. (1987b).

Second, 5-HT produced in the gut enters the systemic blood circulation (figure 3), where most of this 5-HT is rapidly cleared by the liver, the lungs, and other organs (Thomas and Vane, 1967; Anderson et al., 1987b). Some of the remaining 5-HT is taken up and stored by blood platelets that express SERT but that do not synthesize 5-HT themselves. Blood platelets are an essential component of the blood, where they play important roles in blood clotting and in the regulation of vascular tone. Blood platelets are minute (around 1.5 – 3.0 μm in diameter) protoplasmic disks covered with membrane. They split off from long processes of large (up to 60 μm in diameter), polyploid megakaryocytes in the bone marrow. This origin of blood platelets was demonstrated by J.H. Wright in 1906 (Bremer, 1936); one of his original figures is reproduced in figure 4. Blood platelets are short-lived; their half-life has been estimated to be 4 – 6 days (Heyssel, 1961; Stuart et al., 1975).

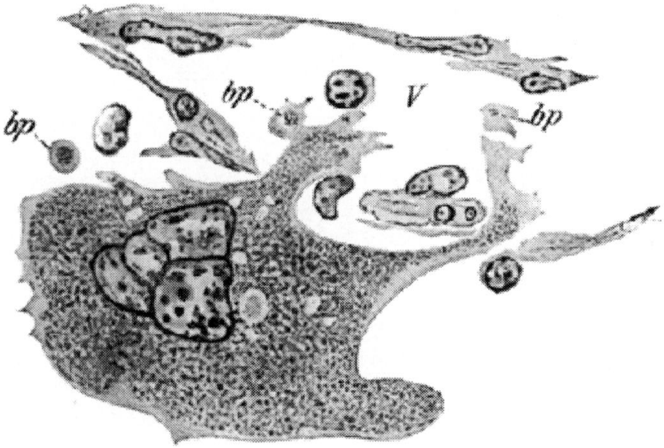

Figure 4. Blood platelet production (bp, blood platelets; V, blood vessel). This original J.H. Wright's figure is reproduced from Bremer (1936).

Since 5-HT is constantly added to and cleared from the systemic blood ciculation, the concentration of *free* 5-HT in the blood plasma varies in different parts of the peripheral 5-HT system (figure 3). In practice, this concentration is usually measured in the distal venous blood, after the liver and the lungs have removed most of the 5-HT released from the gut (figure 3). In the distal venous blood, the concentration of free 5-HT is very low. It has been estimated to be around 304 ng/L (Anderson et al., 1987b) and 0.77 nM (Beck et al., 1993). Because of considerable technical challenges, its actual value remains unknown; however, it is undoubtedly lower than 1000 ng/L (Anderson, 2007). The levels of 5-HT in blood platelets are a few orders of magnitude higher; therefore, whole-blood 5-HT levels are virtually equivalent to the levels of the 5-HT sequestered in blood platelets. In normal subjects, these whole-blood 5-HT levels have been estimated to be around 187 – 297 μg/L (McBride et al., 1998) and 0.4 μM

(Melke et al., 2008); other studies have reported comparable concentrations (Anderson et al., 1990; Coutinho et al., 2004; Mulder et al., 2004; Hranilovic et al., 2007; see also figure 2 of Janušonis (2008)).

It is unlikely that blood platelets can escape CNS blood capillaries and cross the BBB to reach the brain. Even if this were possible before the BBB has developed, platelets would have to rupture inside the brain to release their 5-HT content. Considering that platelets are important in blood clotting, such a process appears highly improbable. However, it is worth noting that blood platelets share many common features with serotonergic synaptosomes (Stahl, 1977); this similarity may be accidental or not.

In the peripheral 5-HT system, 5-HT is converted to 5-HIAA that can be detected in urine samples. Expressed per mg creatinine (in order to normalize for the body surface area and some other variables), urinary 5-HIAA concentration has been estimated to be 3.5 μg/mg creatinine in normal subjects (Minderaa et al., 1987). Urine also contains 5-HT; urinary 5-HT levels have been found to be around 0.1 μg/mg creatinine in normal subjects (Anderson et al., 1989).

8. Platelet Hyperserotonemia of Autism

We now return to the conceptual approach we introduced earlier (figure 1) and ask the first question: What are the molecular mechanisms that underlie the platelet hyperserotonemia of autism?

The causes of this phenomenon remain poorly understood. In an early study, SERT was implicated in autism (Cook et al., 1997). This study offered an insightful approach: blood platelets and neurons express the same SERT gene (Lesch et al., 1993) and, therefore, some polymorphic SERT variants may both overload blood platelets with 5-HT and alter serotonergic signaling in the developing brain. More recent studies have shown that SERT polymorphic variants do partially determine platelet 5-HT uptake rates (Anderson et al., 2002b), but that these polymorphisms, alone, do not cause the platelet hyperserotonemia of autism (Anderson et al., 2002b; Persico et al., 2002) or contribute to autism itself (Persico et al., 2000; Coutinho et al., 2004; Wu et al., 2005; Ramoz et al., 2006).

Other studies have suggested that the platelet hyperserotonemia may be caused by altered 5-HT release from the gut (Minderaa et al., 1987; Croonenberghs et al., 2005). However, currently there is no strong evidence to support this hypothesis. It is likely that altered 5-HT release from the gut, alone, cannot elevate platelet 5-HT levels.

Solving this problem may require more than good experimental approaches. In an attempt to move away from common-sense intuitions (which are often unreliable when a system contains feedback loops), I have recently introduced a quantitative model of the platelet hyperserotonemia of autism (Janušonis, 2005). This model is based on the known physiology of the peripheral 5-HT system, but it also uses additional assumptions.

First, I will briefly summarize the structure of the original model. Second, I will show that some of the potentially weak assumptions of the original model are unnecessary and can be relaxed if a slightly different mathematical approach is taken. Third, I will discuss a new model that further reduces the number of assumptions that have not yet been validated in experimental studies.

9. Blood Platelet 5-HT Levels: Model I

The original model (Janušonis, 2005) is based on the 5-HT circulation in the periphery (figure 3). The gut produces 5-HT that is released into the systemic blood circulation, where most of the 5-HT is rapidly cleared by the liver, lungs, and other organs (Thomas and Vane, 1967; Anderson et al., 1987b). Next, an important assumption is made: when the blood re-enters the gut after one full circulation, the remaining free 5-HT in the blood plasma is detected by gut cells that express 5-HT receptors and that can adjust 5-HT release from the gut. These hypothetical detector-cells may be EC cells that synthesize most of the gut 5-HT and that express 5-HT_{1A} receptors (Kirchgessner et al., 1996). However, the properties of the model would not change if the free 5-HT were detected by other cells (e.g., serotonergic or non-serotonergic neurons of the ENS) that then altered 5-HT release from EC cells.

The model is described by a system of discrete recursive equations:

$$\frac{R_{t+T} - R_{set}}{R_{set}} = \alpha \frac{F_{set} - (1-\gamma)F_t}{F_{set}}, \tag{1}$$

$$F_{t+T} = (1-\gamma)F_t + R_{t+T}, \tag{2}$$

where T is the time in which blood completes one circulation cycle; R_{t+T} is the 5-HT release rate from the gut at time $t+T$; R_{set} is the pre-set ("normal") 5-HT release rate from the gut; F_t is the flow of free 5-HT in the blood as the blood exits the gut at time t; F_{set} is the pre-set ("normal") flow of free 5-HT in the blood that enters the gut; $\alpha \geq 0$ is the adjustment strength ("gain") of the 5-HT release rate from the gut; and γ ($0 < \gamma < 1$) is the proportion of free 5-HT that is removed from the blood in one circulation cycle.

The structure of the model is straightforward. Suppose we follow a small volume of blood that has just left the gut (at time t). The flow of free 5-HT in this volume is expected to be high; it is F_t. Soon after the volume leaves the gut, most of the free 5-HT is cleared by the liver, lungs, and other organs; the remaining flow of free 5-HT is $(1 - \gamma)F_t$. As the blood volume re-enters the gut, this remaining flow is compared to a pre-set value (F_{set}) and the gut 5-HT release rate is adjusted accordingly (R_{t+T}). Negative feedback is assumed; therefore, the 5-HT release rate is increased if the 5-HT flow is too low and decreased if it is too high. Finally, as the blood volume leaves the gut again, its flow of free 5-HT (F_{t+T}) is the sum of the free 5-HT flow *before* the blood entered the gut and the gut 5-HT release rate at time $t + T$.

Next, we assume that the 5-HT concentration in blood platelets is a linear function of the steady-state flow of free 5-HT in the blood after most of the 5-HT released from the gut has been cleared by the liver, lungs, and other organs. The linearity of this relationship can be assumed because the Michaelis-Menten constant for platelet uptake is considerably larger (approximately 0.5 μM) than the levels of free 5-HT (Anderson et al., 1987b; Beck et al., 1993; Anderson et al., 2002b; Anderson, 2007). This relationship can be studied at the steady state because 5-HT uptake by platelets is slow and continues throughout their life span (Mezzano et al., 1984). Less is known about when platelets take up most of their 5-HT. Previous studies have hypothesized, based on physiological data, that platelets take up little 5-HT released from the gut before blood reaches the liver and the lungs (Anderson et al., 1987b). Then platelet 5-HT levels (P) can be expressed as

$$P = A(1-\gamma)\hat{F}, \tag{3}$$

where *F-hat* is the steady-state value of F_t and A is a constant. Since the steady state is defined as $F_t = F_{t+T} \equiv$ *F-hat* and $R_t = R_{t+T} \equiv$ *R-hat*, it follows immediately from equations 1 – 3 that

$$P(\alpha,\gamma) \equiv P = A\frac{F_{set}R_{set}(\alpha+1)(1-\gamma)}{R_{set}\alpha(1-\gamma)+\gamma F_{set}}. \quad (4)$$

It also follows from equations 1 – 3 that

$$\hat{R} = \gamma\hat{F} \quad (5)$$

and that

$$P = -\frac{AF_{set}}{\alpha R_{set}}\hat{R} + AF_{set}\frac{\alpha+1}{\alpha}, \quad (6)$$

assuming $\alpha \neq 0$.

Equation (6) states that *higher* steady-state 5-HT release rates of the gut should correspond to *lower* platelet 5-HT levels, *if* (important!) α is held constant. In other words, if one were to sample a group of individuals with the same α, measuring their platelet 5-HT levels and gut 5-HT release rates precisely, the correlation coefficient between these two variables would be minus one. In practice, such a study would be extremely difficult to carry out, because gut 5-HT release can be measured only indirectly (e.g., as urinary 5-HIAA levels) and α cannot be controlled experimentally or statistically (as a covariate). One study has found no significant correlation between the levels of urinary 5-HIAA and whole blood 5-HT in autistic and control groups (Minderaa et al., 1987). However, the value of the sample correlation was not reported and urinary 5-HIAA levels provide only an indirect measure of how much 5-HT is produced by the gut. Recently, this problem has been investigated in a new experimental study (Mulder et al., 2009). With no evidence to support or invalidate it, equation (6) should be considered a hypothetical prediction. What is perhaps more important is that we reached this counterintuitive conclusion by making simple, common-sense assumptions. This serves as a reminder that we do not have a good intuitive understanding of even simple processes with feedback loops.

When physiologically meaningful values are plugged into equation (4) (Janušonis, 2005), platelet 5-HT levels turn out to depend strongly on the interaction between the factor that controls 5-HT release from the gut (α) and the factor that determines 5-HT clearance after blood leaves the gut (γ). Specifically, a low α is a necessary but not sufficient condition for platelet hypersertonemia to occur. It is only when α is low in the presence of a low γ that blood platelets accumulate unusually high 5-HT levels (figure 5). The actual molecules behind α and γ may be a 5-HT receptor and SERT, respectively, but other functionally similar molecules may play these roles as well. If *only one* of these two molecules plays a key role in the development of the autistic brain (figure 1), this interaction may explain why, in representative and diagnostically homogeneous samples of autistic individuals, many autistic subjects have normal platelet 5-HT levels, even though some other individuals always have abnormally high 5-HT levels. Obviously, the hypothesis that *two different factors determine platelet 5-HT levels but only one of them plays a key role in the developing autistic brain* can be considered independently of the present model.

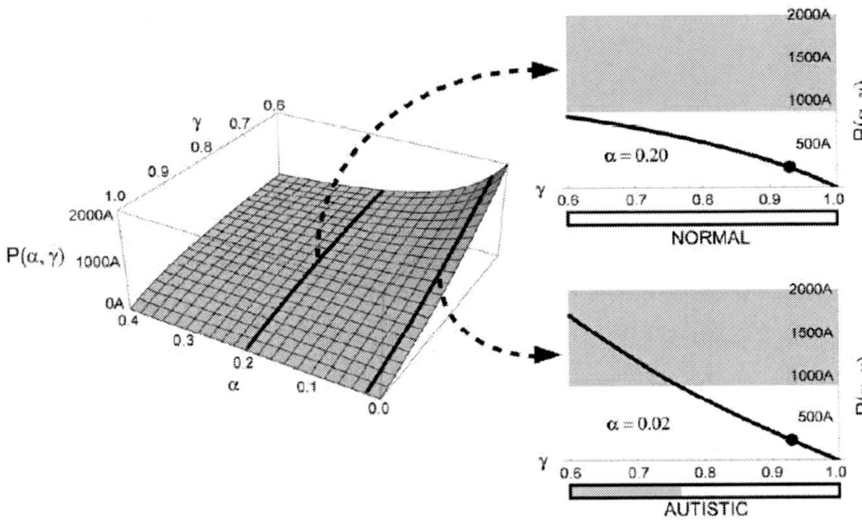

Figure 5. Model I: Platelet levels as a function of the factor that controls 5-HT release from the gut (α) and the factor that clears free 5-HT from the blood circulation (γ). This function is given by equation (4). Note that if α is normal (high), platelet 5-HT levels remain low with any γ, but if α is autistic (low), individuals with a low γ become hyperserotonemic. The black circles mark the points whose coordinates are independent of α and are $\gamma^* = R_{set}/(R_{set} + F_{set})$ and $P(\alpha, \gamma^*) = AF_{set}$. The values of the parameters are F_{set} = 210 ng/min, R_{set} = 3000 ng/min, α = 0.20 (normal) or α = 0.02 (autistic). See Janušonis (2005) for details.

10. Platelet 5-HT Levels: Revised Model I

The presented model is supported by the evidence that EC cells express 5-HT_{1A} receptors (Kirchgessner et al., 1996) that may potentially adjust 5-HT release from the gut depending on extracellular 5-HT levels. Some studies have suggested that EC cells also express 5-HT_2, 5-HT_3 and 5-HT_4 receptors (Gebauer et al., 1993; Racke et al., 1996; Schworer and Ramadori, 1998), but more recent studies have failed to find 5-HT_3 and 5-HT_4 receptors in these cells (Schafermeyer et al., 2004; Liu et al., 2005). Regardless of what 5-HT receptors are actually expressed by EC cells, 5-HT_{1A}, 5-HT_3 and 5-HT_4 receptors are expressed by ENS neurons (Kirchgessner et al., 1992; Tack and Sarnelli, 2002; Gershon and Tack, 2007) that can potentially modulate 5-HT release from the gut.

However, the proposed model has a problem in that it postulates that the gut can detect free 5-HT in the arriving blood, which is a highly hypothetical scenario. It is more likely that gut 5-HT release is controlled not by the levels of free 5-HT in the blood but rather by local 5-HT levels in the gut itself. On the other hand, platelet 5-HT plays important roles in blood clotting and vascular tone and should be homeostatically regulated, even though we may not understand the exact mechanisms. Therefore, we can make a deliberately vague assumption that, *at the steady state*, the gut 5-HT release rate is a function of the platelet 5-HT levels. Importantly, this does not imply that platelet 5-HT directly controls gut 5-HT release. This would be unlikely, considering that platelet 5-HT is isolated from the blood plasma by the platelet membrane. The mechanism can have many steps and can be as complex as one may wish.

Mathematically, this can be stated as follows:

$$\hat{R} = f(P), \tag{7}$$

where f is some analytic function. Next, we simply rewrite equations (2) and (3) at the steady state. These two equations are obviously independent of any assumptions about how gut 5-HT release is regulated:

$$\hat{F} = (1-\gamma)\hat{F} + \hat{R}, \tag{8}$$

$$P = A(1-\gamma)\hat{F}. \tag{9}$$

If P does not deviate much from some pre-set value P_{set}, we can expand f into a Taylor (power) series in the neighborhood of P_{set}:

$$\hat{R} = f(P_{set}) + f'(P_{set})(P - P_{set}) + \frac{1}{2} f''(P - P_{set})^2 + \dots \tag{10}$$

If we drop higher-order terms and define the pre-set 5-HT release rate as the rate at which platelet 5-HT levels are "normal" (i.e., $R_{set} \equiv f(P_{set})$), we obtain

$$\hat{R} = R_{set} + \beta(P_{set} - P), \tag{11}$$

where for brevity we defined $\beta \equiv -f'(P_{set})$.

From equations (8) and (9) we obtain that, on the other hand,

$$\hat{R} = \frac{\gamma P}{A(1-\gamma)}. \tag{12}$$

It follows from equations (11) and (12) that

$$P = \frac{A(R_{set} + \beta P_{set})(1-\gamma)}{A\beta(1-\gamma)+\gamma}. \tag{13}$$

This equation is equivalent to equation (4). It becomes apparent if we introduce, purely formally, two new variables:

$$F_{set} \equiv P_{set}/A \tag{14}$$

and

$$\alpha \equiv \frac{\beta P_{set}}{R_{set}}. \tag{15}$$

This yields

$$P = A\frac{F_{set}R_{set}(\alpha+1)(1-\gamma)}{R_{set}\alpha(1-\gamma)+\gamma F_{set}}, \qquad (16)$$

which is exactly equation (4).

It is worth reiterating that we achieved this by relying on very general ideas about homeostasis and without making any specific assumptions about the exact nature of the regulation of 5-HT release from the gut.

11. Platelet 5-HT Levels: Model II

The revised model is based on fewer assumptions than the original model, but it still fails to address a few important problems.

First, it still assumes that blood platelets take up little 5-HT before most of the 5-HT produced by the gut is cleared by the liver, lungs, and other organs. This assumption is based on some physiological evidence (Anderson et al., 1987b), but its validity has not been systematically explored.

Second, the model assumes that 5-HT clearance from the blood circulation is independent of both gut 5-HT release and platelet 5-HT uptake. This may not be so, because all three processes (5-HT release from the gut, 5-HT clearance from the blood circulation and 5-HT uptake by blood platelets) are controlled, in part, by the same SERT. The gut 5-HT release rate is likely to depend on the activity of SERT, because gut 5-HT appears to enter the blood circulation by simple diffusion. The rate of this diffusion should depend on the concentration of *free* 5-HT in the gut wall, which, in turn, is likely to depend on the activity of SERT (Gershon, 2004). Likewise, 5-HT clearance from the blood circulation depends, at least in part, on SERT expression in various organs. (Despite the importance of SERT, it may not account for all 5-HT clearance in the gut and other organs (Chen et al., 2001)). Finally, SERT is the molecular pump that moves 5-HT molecules into blood platelets. The importance of SERT in this process can be clearly demonstrated in SERT-deficient mice that have negligible blood 5-HT levels (Chen et al., 2001). These multiple roles of SERT pose serious challenges. For instance, high activity of SERT may *reduce* the levels of free 5-HT in the gut wall and, consequently, *reduce* free 5-HT levels in the blood plasma, but this same SERT is likely to *increase* 5-HT uptake by blood platelets. If all of these processes are considered, will blood platelets have more or less 5-HT?

These apparent complications are addressed in the most recent model of the platelet hyperserotonemia of autism that is presented in detail elsewhere (Janušonis, 2008). Regarding the first problem, we should be able to calculate the average 5-HT uptake rate of blood platelets if we know their numerical concentration, life span, and the steady-state concentration of 5-HT in the whole blood. In other words, if we know what 5-HT concentration has to be maintained in the whole blood and we know how many platelets are available, the average uptake rate of a platelet with a known life span becomes completely predetermined. Specifically, it can be shown (Janušonis, 2008) that this rate is

$$u = \frac{C_s}{\tau C_p}, \quad (17)$$

where C_s is the concentration of 5-HT in the whole blood, C_p is the numerical concentration of platelets in the blood and $\tau = 1.44\ t_{1/2}$, where $t_{1/2}$ is the half-life of platelets. In normal humans, C_s/C_p has been experimentally estimated to be approximately $3.58{\cdot}10^{-18}$ mol/platelet (Mulder et al., 2004). The half-life of platelets is approximately 5 days (Heyssel, 1961; Stuart et al., 1975). It follows that $u = 3.44{\cdot}10^{-22}$ mol/min. If Michaelis-Menten kinetics is assumed with $K_m \approx 0.6$ μM (Anderson et al., 1987b, 2002b) and $V_{max} \approx 1.26{\cdot}10^{-18}$ mol/(min·platelet) (Anderson et al., 2002b), the free 5-HT concentration that correspond to this uptake rate is 0.16 nM. This concentration is on the same order as an accurate experimental estimate of free 5-HT in the distal venous plasma obtained by Beck et al. (1993). This strongly suggests that platelets take up little 5-HT before most of the 5-HT released from the gut is removed from the circulation by the liver, lungs, and other organs.

Regarding the second problem, it can be shown (Janušonis, 2008) that, if a number of key components of the peripheral 5-HT system are considered,

$$P = A\hat{R}\frac{k_p(x)}{k_g(x) + a\Theta(x) + bk_g(x)\Theta(x)}, \quad (18)$$

where, again, P is the platelet 5-HT levels; *R-hat* is the steady-state gut 5-HT release rate; x is the activity of SERT; $k_p(x)$, $k_g(x)$ and $\Theta(x)$ are increasing functions of x, representing the platelet 5-HT uptake rate constant, the 5-HT uptake rate constant in the gut wall, and the 5-HT clearance in the liver, lungs, and other organs, respectively; and *A, a,* and *b* are constants. In equation (18), the 5-HT production rate is assumed to be independent of the actual 5-HT levels in the gut wall and blood platelets. This relationship is similar to equations (4) and (16) when $\alpha = 0$. The full version of equation (18) includes regulation of gut 5-HT release, as well as other parameters and their interactions; it yields platelet 5-HT levels that are consistent with values reported in experimental studies (Janušonis, 2008). We are currently testing some predictions of the model experimentally (Albay et al., 2009).

12. From Platelet Hyperserotonemia to the Developing Brain

Models I and II suggest that the platelet hypersertonemia of autism may be caused by an interaction between a factor that determines how much 5-HT is released from the gut and a factor that removes free 5-HT from the blood circulation. We also hypothesized that one of these two factors may play a key role in the early development of the autistic brain, where its function may be different from that in the peripheral 5-HT system (figure 1).

Recent evidence appears to be consistent with these predictions. Platelet 5-HT levels have been shown to depend on the interaction between *Tph1* and SERT polymorphic variants, as well as between the $\beta 3$ integrin subunit and SERT polymorphic variants (Coutinho et al., 2007). Also, altered melatonin levels have been reported in the blood plasma of autistic individuals (Melke et al., 2008). Since plasma melatonin is known to be synthesized in the gut and in the pineal gland (Bubenik et al., 1996; Saha et al., 2007), this change may indicate abnormalities in the gut 5-HT-melatonin synthesis pathway.

In order to test the validity of the general conceptual approach presented here (figure 1), we have analyzed the brain, gut and blood 5-HT levels in developing mice lacking the 5-HT_{1A} receptor. By the third postnatal week, the mean platelet 5-HT levels in the knockout mice were 24% higher than those in age-matched wild-type mice (Janušonis et al., 2006). This increase falls within the range of the typical platelet hyperserotonemia observed in autistic groups (Anderson et al., 1990).

The 5-HT_{1A} receptor is expressed in EC cells (Kirchgessner et al., 1996) and it is also known to have high levels of expression in many areas of the developing rat and human brain (Bar-Peled et al., 1991; Miquel et al., 1994; del Olmo et al.,

1994; del Olmo et al., 1998; Patel and Zhou, 2005). The lack of this receptor during a critical developmental period results in an "anxious" behavioral phenotype in mice (Gross et al., 2002). Interestingly, the Purkinje cells in the rat cerebellum express high levels of the 5-HT_{1A} receptor during development, but their adult expression levels are very low (Miquel et al., 1994). Reduced Purkinje cell numbers have been one of the most consistent observations in autistic brains (Williams et al., 1980; Ritvo et al., 1986; Kern, 2003). Mice lacking 5-HT_{1A} receptors may also have Purkinje cell abnormalities, but the specificity of this phenomenon may be difficult to prove, since many mouse knockouts exhibit Purkinje cell deficits (DiCicco-Bloom et al., 2006). In humans, the density of brain 5-HT_{1A} autoreceptors has been shown to correlate with the reactivity of the amygdala (Fisher et al., 2006).

The findings that 5-HT_{1A} receptor-knockout mice develop a specific behavioral phenotype (Ramboz et al., 1998; Parks et al., 1998; Heisler et al., 1998) *and* postnatal platelet hyperserotonemia (Janušonis et al., 2006) suggest that the development of the autistic brain and the autistic platelet hyperserotonemia may potentially be caused by abnormal function of a 5-HT receptor expressed in the developing brain *and* in the gut. One study has reported a decrease in the expression of 5-HT_{1A} receptors in autistic postmortem brains (Pickett and London, 2005). However, the 5-HT_{1A} receptor remains the only 5-HT receptor studied with respect to platelet hyperserotonemia and it is not known if other 5-HT receptors also contribute to platelet 5-HT levels.

An understanding of the origin of the blood hyperserotonemia of autism may allow us to focus on just a few 5-HT receptors that may set the developing brain on the autistic trajectory. We have shown that early serotonergic afferents arriving in the developing cerebral cortex make synaptic contacts with Cajal-Retzius (CR) cells in embryonic mice (Janušonis et al., 2004). CR cells are transient cells in the cortical marginal zone (the future cortical layer I) that play an important role in the positioning of migrating neurons (Soriano and Del Rio, 2005). Such synaptic contacts are likely to be present in the primate brain as well, since the anatomical distribution of early serotonergic fibers appears to be virtually the same in the mouse, rat and human brain (Wallace and Lauder, 1983; Bruning et al., 1997; Verney et al., 2002) and CR cells have been described in all studied vertebrate species, including humans (Molnar et al., 2006; Cabrera-Socorro et al., 2007; Tissir and Goffinet, 2007). In mammals, CR cells are uniquely characterized by co-expression of the glycoprotein reelin and the transcription factor *p73* (Meyer et al., 2002; Tissir and Goffinet, 2007). In the absence of reelin, the normal positioning of neurons in the developing mouse cortex becomes disrupted (Tissir and Goffinet, 2003).

Very little is known about which 5-HT receptors are expressed in rodent CR cells and nothing is known about 5-HT receptors in human CR and CR-like (Meyer and Goffinet, 1998) cells. Based on previous reports, it has been suggested that CR cells may express 5-HT_{1A} receptors (Janušonis et al., 2004) or 5-HT_3 receptors (Tecott et al., 1995), both of which are also expressed in the gut (Kirchgessner et al., 1996; Gershon, 2003; Gershon and Tack, 2007). A recent report has provided new evidence that mouse CR cells do express 5-HT_3 receptors (Chameau et al., 2006).

Altered function of a 5-HT receptor in CR cells may result in altered secretion of reelin and, consequently, in the abnormal cortical minicolumns that have been reported in autistic brains (Casanova et al., 2002; Casanova et al., 2006). The cortical minicolumn is a fundamental functional unit of the primate cerebral cortex that reflects the radial migration of neurons during development (Casanova et al., 2003). An abnormal structure of cortical minicolumns may radically alter information processing in the cerebral cortex and may lead to a number of autistic symptoms, including increased incidence of seizures and auditory-tactile hypersensitivity (Casanova, 2006). Intriguingly, reelin signaling does appear to be affected in autism (Fatemi et al., 2005; Serajee et al., 2006). However, it remains unclear whether these findings are indicative of altered reelin secretion by fetal CR cells, since CR cells disappear at the end of cortical development and reelin may be expressed by many other brain and peripheral cells. Also, CR cells express MeCP2, a protein that is mutated in Rett syndrome and that may be important in autism (Zoghbi, 2003). Abnormal 5-HT signaling at CR cells may alter the expression of this protein and perturb cortical development.

13. Conclusions

The hyperserotonemia of autism may offer a powerful approach to the early development of the autistic brain that cannot be observed directly. The current state of autism research indicates that a combination of theoretical and experimental approaches, as well as close collaboration between groups working on different 5-HT systems will be necessary to understand the logic of autism spectrum disorders.

14. Acknowledgments

This work was supported, in part, by the Santa Barbara Cottage Hospital-UCSB Special Research Award and UCSB Academic Senate Faculty Research grants. I thank Dr. K.V. Fite for a copy of Bremer (1936) and Vaiva for the encouragement and support.

15. References

Abrahams B.S., Geschwind D.H. (2008) Advances in autism genetics: On the threshold of a new neurobiology. *Nat. Rev. Genet. 9*: 341-355.

Adell A., Celada P., Abellan M.T., Artigas F. (2002) Origin and functional role of the extracellular serotonin in the midbrain raphe nuclei. *Brain Res. Brain Res. Rev. 39*: 154-80.

Albay R., Chen A., Anderson G.M., Tatevosyan M., Janušonis S. (2009) Relationships among body mass, brain mass, gut length, and blood serotonin in young, wild-type mice. *Submitted.*

Amaral D.G., Bauman M.D., Schumann C.M. (2003) The amygdala and autism: implications from non-human primate studies. *Genes Brain Behav. 2*: 295-302.

Amaral D.G., Schumann C.M., Nordahl C.W. (2008) Neuroanatomy of autism. *Trends Neurosci. 31*: 137-145.

Anderson G.M. (2002) Genetics of childhood disorders: XLV. Autism, part 4: serotonin in autism. *J. Am. Acad. Child Adolesc. Psychiatry 41*: 1513-1516.

Anderson G.M. (2007) Measurement of plasma serotonin in autism. *Pediatr. Neurol. 36*: 138-139.

Anderson G.M., Bennett A.J., Weld K.P., Pushkas J.G., Ocame D.M., Higley J.D. (2002a) Serotonin in cisternal cerebrospinal fluid of rhesus monkeys: basal levels and effects of sertraline administration. *Psychopharmacology (Berl) 161*: 95-99.

Anderson G.M., Freedman D.X., Cohen D.J., Volkmar F.R., Hoder E.L., McPhedran P., Minderaa R.B., Hansen C.R., Young J.G. (1987a) Whole blood serotonin in autistic and normal subjects. *J. Child Psychol. Psychiatry 28*: 885-900.

Anderson G.M., Gutknecht L., Cohen D.J., Brailly-Tabard S., Cohen J.H., Ferrari P., Roubertoux P.L., Tordjman S. (2002b) Serotonin transporter promoter variants in autism: functional effects and relationship to platelet hyperserotonemia. *Mol. Psychiatry 7*: 831-836.

Anderson G.M., Horne W.C., Chatterjee D., Cohen D.J. (1990) The hyperserotonemia of autism. *Ann. N. Y. Acad. Sci. 600*: 331-340.

Anderson G.M., Minderaa R.B., Cho S.C., Volkmar F.R., Cohen D.J. (1989) The issue of hyperserotonemia and platelet serotonin exposure: a preliminary study. *J. Autism Dev. Disord. 19*: 349-351.

Anderson G.M., Ross D.L., Klykylo W., Feibel F.C., Cohen D.J. (1988) Cerebrospinal fluid indoleacetic acid in autistic subjects. *J. Autism Dev. Disord. 18*: 259-262.

Anderson G.M., Stevenson J.M., Cohen D.J. (1987b) Steady-state model for plasma free and platelet serotonin in man. *Life Sci. 41*: 1777-1785.

Babloyantz A. (1986) *Molecules, Dynamics, and Life: An Introduction to Self-Organization of Matter*. Wiley-Interscience; 1986; pp. 1-345.

Bar-Peled O., Gross-Isseroff R., Ben Hur H., Hoskins I., Groner Y., Biegon A. (1991) Fetal human brain exhibits a prenatal peak in the density of serotonin 5-HT1A receptors. *Neurosci Lett. 127*: 173-176.

Beck O., Wallen N.H., Broijersen A., Larsson P.T., Hjemdahl P. (1993) On the accurate determination of serotonin in human plasma. *Biochem. Biophys. Res. Commun. 196*: 260-266.

Bell J.E., Fryer A.A., Collins M., Marshall T., Jones P.W., Strange R., Hume R. (1991) Developmental profile of plasma proteins in human fetal cerebrospinal fluid and blood. Neuropathol. *Appl. Neurobiol. 17*: 441-456.

Belmonte M.K., Allen G., Beckel-Mitchener A., Boulanger L.M., Carper R.A., Webb S.J. (2004) Autism and abnormal development of brain connectivity. *J. Neurosci. 24*: 9228-9231.

Bischoff S.C., Mailer R., Pabst O., Weier G., Sedlik W., Li Z., Chen J.J., Murphy D.L., Gershon M.D. (2009) Role of serotonin in intestinal inflammation: knockout of serotonin reuptake transporter exacerbates 2,4,6-trinitrobenzene sulfonic acid colitis in mice. *Am. J. Physiol. Gastrointest. Liver Physiol.*, in press.

Bradley S.R., Pieribone V.A., Wang W., Severson C.A., Jacobs R.A., Richerson G.B. (2002) Chemosensitive serotonergic neurons are closely associated with large medullary arteries. *Nat. Neurosci. 5*: 401-402.

Bremer, J.L. *A Text-Book of Histology*. Fifth Ed. Philadelphia, PA: P. Blakiston's Son & Co., Inc.; 1936; p. 221.

Bruning G., Liangos O., Baumgarten H.G. (1997) Prenatal development of the serotonin transporter in mouse brain. *Cell Tissue Res. 289*: 211-221.

Bubenik G.A., Pang S.F., Hacker R.R., Smith P.S. (1996) Melatonin concentrations in serum and tissues of porcine gastrointestinal tract and their relationship to the intake and passage of food. *J. Pineal Res. 21*: 251-256.

Buznikov G.A. (1984) The action of neurotransmitters and related substances on early embryogenesis. *Pharmacol. Ther. 25*: 23-59.

Buznikov G.A., Lambert H.W., Lauder J.M. (2001) Serotonin and serotonin-like substances as regulators of early embryogenesis and morphogenesis. *Cell Tissue Res. 305*: 177-186.

Buznikov G.A., Peterson R.E., Nikitina L.A., Bezuglov V.V., Lauder J.M. (2005) The pre-nervous serotonergic system of developing sea urchin embryos and larvae: pharmacologic and immunocytochemical evidence. *Neurochem Res. 30*: 825-837.

Cabrera-Socorro A., Hernandez-Acosta N.C., Gonzalez-Gomez M., Meyer G. (2007) Comparative aspects of *p73* and Reelin expression in Cajal-Retzius cells and the cortical hem in lizard, mouse and human. *Brain Res. 1132*: 59-70.

Casanova M.F. (2006) Neuropathological and genetic findings in autism: the significance of a putative minicolumnopathy. *Neuroscientist 12*: 435-441.

Casanova M.F. (2007) The neuropathology of autism. *Brain Pathol. 17*: 422-433.

Casanova M.F., Buxhoeveden D., Gomez J. (2003) Disruption in the inhibitory architecture of the cell minicolumn: implications for autism. *Neuroscientist 9*: 496-507.

Casanova M.F., Buxhoeveden D.P., Switala A.E., Roy E. (2002) Minicolumnar pathology in autism. *Neurology 58*: 428-432.

Casanova M.F., van Kooten I.A., Switala A.E., van Engeland H., Heinsen H., Steinbusch H.W., Hof P.R., Trippe J., Stone J., Schmitz C. (2006) Minicolumnar abnormalities in autism. *Acta Neuropathol. (Berl) 112*: 287-303.

Chameau P., Inta D.I., Vitalis T., Monyer H., van Hooft J.A. (2006) Serotonergic control of postnatal cortical development by 5-HT_3 receptors. *Society for Neuroscience Abstracts 125.10*.

Chen J.J., Li Z., Pan H., Murphy D.L., Tamir H., Koepsell H., Gershon M.D. (2001) Maintenance of serotonin in the intestinal mucosa and ganglia of mice that lack the high-affinity serotonin transporter: Abnormal intestinal motility and the expression of cation transporters. *J. Neurosci. 21*: 6348-6361.

Cook E.H., Jr. (1996) Brief report: pathophysiology of autism: neurochemistry. *J. Autism Dev. Disord. 26*: 221-225.

Cook E.H., Jr., Courchesne R., Lord C., Cox N.J., Yan S., Lincoln A., Haas R.,

Courchesne E., Leventhal B.L. (1997) Evidence of linkage between the serotonin transporter and autistic disorder. *Mol. Psychiatry 2*: 247-250.

Courchesne E., Pierce K., Schumann C.M., Redcay E., Buckwalter J.A., Kennedy D.P., Morgan J. (2007) Mapping early brain abnormalities in autism. *Neuron 56*: 399-413.

Courchesne E., Redcay E., Kennedy D.P. (2004) The autistic brain: birth through adulthood. *Curr. Opin. Neurol. 17*: 489-496.

Courchesne E., Redcay E., Morgan J.T., Kennedy D.P. (2005) Autism at the beginning: microstructural and growth abnormalities underlying the cognitive and behavioral phenotype of autism. *Dev. Psychopathol. 17*: 577-597.

Coutinho A.M., Oliveira G., Morgadinho T., Fesel C., Macedo T.R., Bento C., Marques C., Ataide A., Miguel T., Borges L., Vicente A.M. (2004) Variants of the serotonin transporter gene (*SLC6A4*) significantly contribute to hyperserotonemia in autism. *Mol. Psychiatry* 9: 264-271.

Coutinho A.M., Sousa I., Martins M., Correia C., Morgadinho T., Bento C., Marques C., Ataide A., Miguel T.S., Moore J.H., Oliveira G., Vicente A.M. (2007) Evidence for epistasis between *SLC6A4* and *ITGB3* in autism etiology and in the determination of platelet serotonin levels. *Hum. Genet. 121*: 243-256.

Croonenberghs J., Verkerk R., Scharpe S., Deboutte D., Maes M. (2005) Serotonergic disturbances in autistic disorder: L-5-hydroxytryptophan administration to autistic youngsters increases the blood concentrations of serotonin in patients but not in controls. *Life Sci. 76*: 2171-2183.

Dapretto M., Davies M.S., Pfeifer J.H., Scott A.A., Sigman M., Bookheimer S.Y., Iacoboni M. (2006) Understanding emotions in others: mirror neuron dysfunction in children with autism spectrum disorders. *Nat. Neurosci. 9*: 28-30.

del Olmo E., Diaz A., Guirao-Pineyro M., del Arco C., Pascual J., Pazos A. (1994) Transient localization of 5-HT1A receptors in human cerebellum during development. *Neurosci Lett. 166*: 149-152.

del Olmo E., Lopez-Gimenez J.F., Vilaro M.T., Mengod G., Palacios J.M., Pazos A. (1998) Early localization of mRNA coding for 5-HT1A receptors in human brain during development. *Brain Res. Mol. Brain Res. 60*: 123-126.

DiCicco-Bloom E., Lord C., Zwaigenbaum L., Courchesne E., Dager S.R., Schmitz C., Schultz R.T., Crawley J., Young L.J. (2006) The developmental neurobiology of autism spectrum disorder. *J. Neurosci. 26*: 6897-6906.

Duvernoy H.M., Delon S., Vannson J.L. (1981) Cortical blood vessels of the human brain. *Brain Res. Bull. 7*: 519-579.

Fatemi S.H., Snow A.V., Stary J.M., Araghi-Niknam M., Reutiman T.J., Lee S., Brooks A.I., Pearce D.A. (2005) Reelin signaling is impaired in autism. *Biol. Psychiatry 57*: 777-787.

Fisher P.M., Meltzer C.C., Ziolko S.K., Price J.C., Hariri A.R. (2006) Capacity for 5-HT1A-mediated autoregulation predicts amygdala reactivity. *Nat. Neurosci. 9*: 1362-1363.

Ge S., Song L., Pachter J.S. (2005) Where is the blood-brain barrier ... really? *J. Neurosci. Res. 79*: 421-427.

Gebauer A., Merger M., Kilbinger H. (1993) Modulation by 5-HT3 and 5-HT4 receptors of the release of 5-hydroxytryptamine from the guinea-pig small intestine. *Naunyn Schmiedebergs Arch. Pharmacol. 347*: 137-140.

Gershon M.D. (2003) Plasticity in serotonin control mechanisms in the gut. *Curr. Opin. Pharmacol. 3*: 600-607.

Gershon M.D. (2004) Review article: Serotonin receptors and transporters -- roles in normal and abnormal gastrointestinal motility. *Aliment Pharmacol. Ther. 20 Suppl. 7*: 3-14.

Gershon M.D., Tack J. (2007) The serotonin signaling system: From basic understanding to drug development for functional GI disorders. *Gastroenterology 132*: 397-414.

Geschwind D. (2007) Autism: Searching for coherence. *Biol. Psychiatry 62*: 949-950.

Geschwind D.H. (2008) Autism: Many genes, common pathways? *Cell 135*: 391-395.

Gross C., Zhuang X., Stark K., Ramboz S., Oosting R., Kirby L., Santarelli L., Beck S., Hen R. (2002) Serotonin1A receptor acts during development to establish normal anxiety-like behaviour in the adult. *Nature 416*: 396-400.

Hanley H.G., Stahl S.M., Freedman D.X. (1977) Hyperserotonemia and amine metabolites in autistic and retarded children. *Arch. Gen. Psychiatry 34*: 521-531.

Heisler L.K., Chu H.M., Brennan T.J., Danao J.A., Bajwa P., Parsons L.H., Tecott L.H. (1998) Elevated anxiety and antidepressant-like responses in serotonin 5-HT1A receptor mutant mice. *Proc. Natl. Acad. Sci. U.S.A. 95*: 15049-15054.

Heyssel R.M. (1961) Determination of human platelet survival utilizing C-14-labeled serotonin. *J. Clin. Invest. 40*: 2134-2142.

Hranilovic D., Bujas-Petkovic Z., Vragovic R., Vuk T., Hock K., Jernej B. (2007) Hyperserotonemia in adults with autistic disorder. *J. Autism Dev. Disord. 37*: 1934-1940.

Hranilović D., Novak R., Babić M., Novokmet M., Bujas-Petković Z., Jernej B. (2008) Hyperserotonemia of autism: The potential role of 5-HT-related gene variants. *Coll. Antropol. 32 Suppl. 1*: 75-80.

Janušonis S. (2005) Statistical distribution of blood serotonin as a predictor of early autistic brain abnormalities. *Theor. Biol. Med. Model. 2*: 27.

Janušonis S. (2008) Origin of the blood hyperserotonemia of autism. *Theor. Biol. Med. Model. 5*: 10.

Janušonis S., Anderson G.M., Shifrovich I., Rakic P. (2006) Ontogeny of brain and blood serotonin levels in 5-HT_{1A} receptor knockout mice: potential relevance to the neurobiology of autism. *J. Neurochem. 99*: 1019-1031.

Janušonis S., Gluncic V., Rakic P. (2004) Early serotonergic projections to Cajal-Retzius cells: relevance for cortical development. *J. Neurosci. 24*: 1652-1659.

Kemper T.L., Bauman M.L. (2002) Neuropathology of infantile autism. *Mol. Psychiatry 7 Suppl. 2*: S12-S13.

Kern J.K. (2003) Purkinje cell vulnerability and autism: a possible etiological connection. *Brain Dev. 25*: 377-382.

Kirchgessner A.L., Liu M.T., Raymond J.R., Gershon M.D. (1996) Identification of cells that express 5-hydroxytryptamine1A receptors in the nervous systems of the bowel and pancreas. *J. Comp. Neurol. 364*: 439-455.

Lebrand C., Cases O., Adelbrecht C., Doye A., Alvarez C., El Mestikawy S., Seif I., Gaspar P. (1996) Transient uptake and storage of serotonin in developing thalamic neurons. *Neuron 17*: 823-835.

Lebrand C., Cases O., Wehrle R., Blakely R.D., Edwards R.H., Gaspar P. (1998) Transient developmental expression of monoamine transporters in the rodent forebrain. *J. Comp. Neurol 401*: 506-524.

Lebrand C., Gaspar P., Nicolas D., Hornung J.P. (2006) Transitory uptake of serotonin in the developing sensory pathways of the common marmoset. *J. Comp. Neurol. 499*: 677-689.

Lesch K.P., Wolozin B.L., Murphy D.L., Reiderer P. (1993) Primary structure of the human platelet serotonin uptake site: identity with the brain serotonin transporter. *J. Neurochem. 60*: 2319-2322.

Levin M., Buznikov G.A., Lauder J.M. (2006) Of minds and embryos: Left-right asymmetry and the serotonergic controls of pre-neural morphogenesis. *Dev. Neurosci. 28*: 171-185.

Lintas C., Persico A.M. (2008) Autistic phenotypes and genetic testing: State-of-the-art for the clinical geneticist. *J. Med. Genet.*, in press.

Liu M., Geddis M.S., Wen Y., Setlik W., Gershon M.D. (2005) Expression and function of 5-HT4 receptors in the mouse enteric nervous system. *Am. J.*

Physiol. Gastrointest. Liver Physiol. 289: G1148-G1163.

McBride P.A., Anderson G.M., Hertzig M.E., Snow M.E., Thompson S.M., Khait V.D., Shapiro T., Cohen D.J. (1998) Effects of diagnosis, race, and puberty on platelet serotonin levels in autism and mental retardation. *J. Am. Acad. Child Adolesc. Psychiatry 37*: 767-776.

Melke J., Goubran B.H., Chaste P., Betancur C., Nygren G., Anckarsater H., Rastam M., Stahlberg O., Gillberg I.C., Delorme R., Chabane N., Mouren-Simeoni M.C., Fauchereau F., Durand C.M., Chevalier F., Drouot X., Collet C., Launay J.M., Leboyer M., Gillberg C., Bourgeron T. (2008) Abnormal melatonin synthesis in autism spectrum disorders. *Mol. Psychiatry 13*: 90-98.

Meyer G., Goffinet A.M. (1998) Prenatal development of reelin-immunoreactive neurons in the human neocortex. *J. Comp. Neurol. 397*: 29-40.

Meyer G., Perez-Garcia C.G., Abraham H., Caput D. (2002) Expression of *p73* and Reelin in the developing human cortex. *J. Neurosci. 22*: 4973-4986.

Mezzano D., Aranda E., Rodriguez S., Foradori A., Lira P. (1984) Increase in density and accumulation of serotonin by aging human platelets. *Am. J. Hematol.* 17: 11-21.

Minderaa R.B., Anderson G.M., Volkmar F.R., Akkerhuis G.W., Cohen D.J. (1987). Urinary 5-hydroxyindoleacetic acid and whole blood serotonin and tryptophan in autistic and normal subjects. *Biol. Psychiatry 22*: 933-940.

Miquel M.C., Kia H.K., Boni C., Doucet E., Daval G., Matthiessen L., Hamon M., Verge D. (1994) Postnatal development and localization of 5-HT1A receptor mRNA in rat forebrain and cerebellum. *Brain Res. Dev. Brain Res.* 80: 149-157.

Mollgard K., Saunders N.R. (1986) The development of the human blood-brain barrier and blood-CSF barriers. *Neuropathol. Appl. Neurobiol. 12*: 337-358.

Molnar Z., Metin C., Stoykova A., Tarabykin V., Price D.J., Francis F., Meyer G., Dehay C., Kennedy H. (2006) Comparative aspects of cerebral cortical development. *Eur. J. Neurosci. 23*: 921-934.

Mulder E.J., Anderson G.M., Kema I.P., de Bildt A., van Lang N.D., den Boer J.A., Minderaa R.B. (2004) Platelet serotonin levels in pervasive developmental disorders and mental retardation: Diagnostic group differences, within-group distribution, and behavioral correlates. *J. Am. Acad. Child Adolesc. Psychiatry 43*: 491-499.

Mulder E.J., Anderson G.M., Kemperman R.F.J., Oosterloo-Duinkerken A., Minderaa R.B., Kema I.P. (2009) Urinary excretion of 5-hydroxyindoleacetic acid, serotonin and 6-sulphatoxymelatonin in normoserotonemic and hyperserotonemic autistic individuals. *Submitted.*

Nakamura K., Sugawara Y., Sawabe K., Ohashi A., Tsurui H., Xiu Y., Ohtsuji M., Lin Q.S., Nishimura H., Hasegawa H., Hirose S. (2006) Late developmental stage-specific role of tryptophan hydroxylase 1 in brain serotonin levels. *J. Neurosci. 26*: 530-534.

Narayan M., Srinath S., Anderson G.M., Meundi D.B. (1993) Cerebrospinal fluid levels of homovanillic acid and 5-hydroxyindoleacetic acid in autism. *Biol. Psychiatry 33*: 630-635.

Nicolis G., Prigogine I. (1989) *Exploring Complexity: An Introduction*. W.H. Freeman & Company; 1989; pp. 1-328.

Nolte J. (2002) The Human Brain: An Introduction to Its Functional Anatomy. Fifth Ed. St. Louis, MO: Mosby, Inc.; 2002; p. 55.

Palmen S.J., van Engeland H., Hof P.R., Schmitz C. (2004) Neuropathological findings in autism. *Brain 127 (Pt. 12)*: 2572-2583.

Papadopoulos G.C., Parnavelas J.G., Buijs R. (1987) Monoaminergic fibers form conventional synapses in the cerebral cortex. *Neurosci. Lett. 76*: 275-279.

Parks C.L., Robinson P.S., Sibille E., Shenk T., Toth M. (1998) Increased anxiety of mice lacking the serotonin1A receptor. *Proc. Natl. Acad. Sci. U.S.A. 95*: 10734-10739.

Parnavelas J.G., Papadopoulos G.C. (1989) The monoaminergic innervation of the cerebral cortex is not diffuse and nonspecific. *Trends Neurosci. 12*: 315-319.

Patel T.D., Zhou F.C. (2005) Ontogeny of 5-HT1A receptor expression in the developing hippocampus. *Brain Res. Dev. Brain Res. 157*: 42-57.

Persico A.M., Militerni R., Bravaccio C., Schneider C., Melmed R., Conciatori M., Damiani V., Baldi A., Keller F. (2000) Lack of association between serotonin transporter gene promoter variants and autistic disorder in two ethnically distinct samples. *Am. J. Med. Genet. 96*: 123-127.

Persico A.M., Pascucci T., Puglisi-Allegra S., Militerni R., Bravaccio C., Schneider C., Melmed R., Trillo S., Montecchi F., Palermo M., Rabinowitz D., Reichelt K.-L., Conciatori M., Marino R., Keller F. (2002) Serotonin transporter gene promoter variants do not explain the hyperserotonemia in autistic children. *Mol. Psychiatry 7*: 795-800.

Pickett J., London E. (2005) The neuropathology of autism: a review. *J. Neuropathol. Exp. Neurol. 64*: 925-935.

Racke K., Reimann A., Schworer H., Kilbinger H. (1996) Regulation of 5-HT release from enterochromaffin cells. *Behav. Brain Res.* 73: 83-87.

Ramboz S., Oosting R., Amara D.A., Kung H.F., Blier P., Mendelsohn M., Mann J.J., Brunner D., Hen R. (1998) Serotonin receptor 1A knockout: an animal model of anxiety-related disorder. *Proc. Natl. Acad. Sci. U.S.A. 95*: 14476-14481.

Ramoz N., Reichert J.G., Corwin T.E., Smith C.J., Silverman J.M., Hollander E., Buxbaum J.D. (2006) Lack of evidence for association of the serotonin transporter gene *SLC6A4* with autism. *Biol. Psychiatry* 60: 186-191.

Raymond G.V., Bauman M.L., Kemper T.L. (1996) Hippocampus in autism: a Golgi analysis. *Acta Neuropathol. (Berl) 91*: 117-119.

Ritvo E.R., Freeman B.J., Scheibel A.B., Duong T., Robinson H., Guthrie D., Ritvo A. (1986) Lower Purkinje cell counts in the cerebella of four autistic subjects: initial findings of the UCLA-NSAC Autopsy Research Report. *Am. J. Psychiatry 143*: 862-866.

Rodier P.M. (2002) Converging evidence for brain stem injury in autism. *Dev. Psychopathol.* 14: 537-557.

Rodier P.M., Ingram J.L., Tisdale B., Nelson S., Romano J. (1996) Embryological origin for autism: developmental anomalies of the cranial nerve motor nuclei. *J. Comp. Neurol. 370*: 247-261.

Rohen J.W., Yokochi C., Lütjen-Drecoll E. (2006) *Color Atlas of Anatomy: A Photographic Study of the Human Body*. Sixth Ed. Philadelphia, PA: Lippincott Williams & Wilkins; 2006; pp. 132, 291.

Saha L., Malhotra S., Rana S., Bhasin D., Pandhi P. (2007) A preliminary study of melatonin in irritable bowel syndrome. *J. Clin. Gastroenterol. 41*: 29-32.

Schafermeyer A., Gratzl M., Rad R., Dossumbekova A., Sachs G., Prinz C. (2004) Isolation and receptor profiling of ileal enterochromaffin cells. *Acta Physiol. Scand. 182*: 53-62.

Schain R.J., Freedman D.X. (1961) Studies on 5-hydroxyindole metabolism in autistic and other mentally retarded children. *J. Pediatr. 58*: 315-320.

Schumann C.M., Hamstra J., Goodlin-Jones B.L., Lotspeich L.J., Kwon H., Buonocore M.H., Lammers C.R., Reiss A.L., Amaral D.G. (2004) The amygdala is enlarged in children but not adolescents with autism; the hippocampus is enlarged at all ages. *J. Neurosci. 24*: 6392-6401.

Schworer H., Ramadori G. (1998) Autoreceptors can modulate 5-hydroxytryptamine release from porcine and human small intestine in vitro. *Naunyn Schmiedebergs Arch. Pharmacol. 357*: 548-552.

Sebat J., Lakshmi B., Malhotra D., Troge J., Lese-Martin C., Walsh T., Yamrom B., Yoon S., Krasnitz A., Kendall J., Leotta A., Pai D., Zhang R., Lee Y.H., Hicks J., Spence S.J., Lee A.T., Puura K., Lehtimäki T., Ledbetter D., Gregersen P.K., Bregman J., Sutcliffe J.S., Jobanputra V., Chung W., Warburton D., King M.C., Skuse D., Geschwind D.H., Gilliam T.C., Ye K.,

Wigler M. (2007) Strong association of *de novo* copy number mutations with autism. *Science 316*: 445-449.

Serajee F.J., Zhong H., Mahbubul Huq A.H. (2006) Association of Reelin gene polymorphisms with autism. *Genomics 87*: 75-83.

Severson C.A., Wang W., Pieribone V.A., Dohle C.I., Richerson G.B. (2003) Midbrain serotonergic neurons are central pH chemoreceptors. *Nat. Neurosci. 6*: 1139-1140.

Sharma H.S. (2004a) Blood-brain and spinal cord barriers in stress. In: Sharma H.S., Westman, J., eds. *Blood-Spinal Cord and Brain Barriers in Health and Disease*. San Diego, California: Elsevier Academic Press; 2004; pp. 231-298.

Sharma H.S. (2004b) Influence of serotonin on the blood-brain and the blood-spinal cord barriers. In: Sharma H.S., Westman, J., eds. *Blood-Spinal Cord and Brain Barriers in Health and Disease*. San Diego, CA: Elsevier Academic Press; 2004; pp. 117-157.

Soriano E., Del Rio J.A. (2005) The cells of Cajal-Retzius: still a mystery one century after. *Neuron 46*: 389-394.

Stahl S.M. (1977) The human platelet. A diagnostic and research tool for the study of biogenic amines in psychiatric and neurologic disorders. *Arch. Gen. Psychiatry 34*: 509-516.

Stuart M.J., Murphy S., Oski F.A. (1975) A simple nonradioisotope technic for the determination of platelet life-span. *N. Engl. J. Med. 292*: 1310-1313.

Szatmari P. et al. (The Autism Genome Project Consortium) (2007) Mapping autism risk loci using genetic linkage and chromosomal rearrangements. *Nat. Genet. 39*: 319-328.

Tack J., Sarnelli G. (2002) Serotonergic modulation of visceral sensation: upper gastrointestinal tract. *Gut 51 Suppl. 1*: i77-i80.

Thomas D.P., Vane J.R. (1967) 5-hydroxytryptamine in the circulation of the dog. *Nature 216*: 335-338.

Tissir F., Goffinet A.M. (2003) Reelin and brain development. *Nat. Rev. Neurosci. 4*: 496-505.

Tissir F., Goffinet A.M. (2007) Reelin, Cajal-Retzius cells, and cortical evolution. In: Kaas J.H., ed. *Evolution of Nervous Systems: A Comprehensive Reference*. San Diego, CA: Elsevier, Academic Press; 2007; pp. 89-97.

Vargas D.L., Nascimbene C., Krishnan C., Zimmerman A.W., Pardo C.A. (2005) Neuroglial activation and neuroinflammation in the brain of patients with autism. *Ann. Neurol. 57*: 67-81.

Virgintino D., Errede M., Robertson D., Capobianco C., Girolamo F., Vimercati A., Bertossi M., Roncali L. (2004) Immunolocalization of tight junction proteins in the adult and developing human brain. *Histochem. Cell. Biol. 122*: 51-59.

Verney C., Lebrand C., Gaspar P. (2002) Changing distribution of monoaminergic markers in the developing human cerebral cortex with special emphasis on the serotonin transporter. *Anat. Rec. 267*: 87-93.

Wallace JA, Lauder JM. (1983) Development of the serotonergic system in the rat embryo: an immunocytochemical study. *Brain Res. Bull. 10*: 459-479.

Walther DJ, Bader M. (2003) A unique central tryptophan hydroxylase isoform. *Biochem. Pharmacol.* 66(9):1673-80.

Walther D.J., Peter J.U., Bashammakh S., Hortnagl H., Voits M., Fink H., Bader M. (2003) Synthesis of serotonin by a second tryptophan hydroxylase isoform. *Science 299*: 76.

Wenzel D., Felgenhauer K. (1976) The development of the blood-CSF barrier after birth. *Neuropadiatrie. 7*: 175-181.

Whitaker-Azmitia PM. (2001) Serotonin and brain development: role in human developmental diseases. *Brain Res. Bull. 56*: 479-485.

Williams R.S., Hauser S.L., Purpura D.P., DeLong G.R., Swisher C.N. (1980) Autism and mental retardation: neuropathologic studies performed in four retarded persons with autistic behavior. *Arch. Neurol. 37*: 749-753.

Wu S., Guo Y., Jia M., Ruan Y., Shuang M., Liu J., Gong X., Zhang Y., Yang J., Yang X., Zhang D. (2005). Lack of evidence for association between the serotonin transporter gene (*SLC6A4*) polymorphisms and autism in the Chinese trios. *Neurosci Lett. 381*: 1-5.

Zoghbi H.Y. (2003) Postnatal neurodevelopmental disorders: meeting at the synapse? *Science 302*: 826-830.

In: Causes and Risks for Autism
Editors: A. C. Giordano et al.

ISBN: 978-1-60456-861-5

Chapter II

Autism Spectrum Disorders: From Candidate Genes to Candidate Ontology Terms

Valentino Romano**[*1,4]**, Claudia Coronnello**[2]**, Salvatore Micciché**[2]**,
***Silvia Sbacchi**[1]** and Rosario Nunzio Mantegna**[2]*
[1] Dipartimento di Oncologia Sperimentale e Applicazioni Cliniche
[2] Dipartimento di Fisica e Tecnologie Relative;
Università degli Studi di Palermo, Palermo;
[4]Associazione Oasi Maria SS. (I.R.C.C.S.), Troina, Italy

Abstract

Genetic heterogeneity for a multifactorial disease such as autism would imply that any two patients are unlikely to share the same susceptibility loci. Since different susceptibility loci may affect the same function and since functions may be considered at different levels of analysis, the following question arises: at what phenotypic levels is convergence attained by different genes in autism? This is an important question to answer in order to shed light on subcellular, cellular or multicellular structures and functions

[*] Corresponding author: Valentino ROMANO; Dipartimento di Oncologia Sperimentale e Applicazioni Cliniche, Università degli Studi di Palermo, Via S. Lorenzo Colli, 312 Palermo, Italy; Phone: ++3909166722852 ; e-mail: vromano@unipa.it

possibly involved in the pathogenesis of autism. Among the various attempts made to answer this question it is worth mentioning the use of the best available genetic information to focus on the formation, structure and function of the glutamatergic synapse. How could we use this working scheme in a more efficient and hopefully fruitful way?

The recent advent of genomic technology offers an unprecedented opportunity to study the biological structure of autism deep into its molecular roots. For example, recent work on structural genomics has led to the identification of many Copy Number Variations in the gene-rich regions of the genome detected in a large number of patients. Interesting advances have also been made by analyzing the global gene-expression profiles in blood cells of autistic patients. The large amount of data generated by these studies is likely to contain precious information about the pathogenesis of autism. In this paper we review the literature on the genomic studies in ASD and present the results of a preliminary study showing how candidate biological processes for ASD can be identified from microarray data using the Gene Ontology database. We also discuss the importance of the transition from a strategy based on the search for candidate genes to the search for candidate ontology terms to expand our current understanding on the biological processes impaired in ASD.

Introduction

In the last 25–30 years many research efforts have been made to investigate the biological bases of autism, but still early diagnosis, prevention and effective treatments for this disorder appear as long-term goals. At the same time, and reassuringly, significant progress has been made in many specialized areas of autism research, from genetics to cognitive neurosciences (Frith, 2003). Perhaps one of the main contributions from these specialized studies has been the realization that the monolithic autism phenotype can be destructured and studied at several different levels of analysis. Symptomatic of this paradigm shift has been the introduction of the important concept of "endophenotype" (Flint and Munafo, 2007). MRI, neuropathological and immunological findings, macrocephaly, the alteration of certain metabolites in blood or cerebral spinal fluid are all examples of endophenotypes that have been detected in autistic patients. In principle, studying each of these and other endophenotypes offers the opportunity to open a window on the actual molecular, cellular and physiological processes that may have been impaired by genetic and/or environmental factors during critical periods of the Central Nervous System (CNS) development. With more and more data coming out from these specialized fields it is becoming increasingly clear how important is to integrate this information to conceptualize realistic models of autism pathogenesis. To bridge the information generated at different levels of analysis is a necessary step before addressing the more ambitious goal of linking genes to the actual clinical phenotype observed in patients. In this context, it may be useful to recall the simple, but crucial notion that genes "encode" proteins and not repetitive behaviour or theory of mind. These concepts find support in the important advances that have been made integrating molecular data with sub-cellular, cellular or multicellular processes (e.g. neuroligin genes and synaptogenesis; Jamain et al., 2003). These achievements have been mirrored by an analogous progress made integrating, for example, psychological/behavioral data with MRI data (e.g., Bird et al., 2004; Koshino et al., 2007).

It has been proposed that behavioral and cognitive impairments in autistic patients may result from an abnormal function in complex networks of neurons generating a skewed balance of local versus long-range functional connectivity (Belmonte et al., 2004a). More specifically, these impairments would result from an abnormally low activity of brain regions subserving social cognition, along with abnormally high activity of regions subserving lower-level, perceptual processing (Belmonte et al., 2004b). This theory is very important because it offers an excellent opportunity of integrating different types of neurobiological data (including genetic ones) and to formulate fruitful hypotheses. One such hypothesis proposes that predisposing genetic factors are likely to exert their effect on the structure and dynamics of neural systems, that is, abnormal expression within complex networks of genes translates into abnormal function in complex networks of neurons (Belmonte and Bourgeron, 2006). For example, the balance between excitatory and inhibitory neural processes depends on the expression of many genes directly or indirectly involved in neurotransmission, including receptors, membrane transporters, enzymes for neurotransmitter synthesis or degradation, cytoskeletal, and vescicular proteins, signaling and effector proteins, and regulators of transcription and translation (Belmonte and Bourgeron, 2006). On the other hand, since predisposing genetic factors are likely to exert an ongoing action at many if not all phases of CNS development, the molecular and cellular mechanisms involved in the early, middle and late phases of brain development, should also be taken into account and investigated. For instance, a decreased apoptosis has been proposed to explain various neuropathological findings in autistic patients (Saskia et al., 2004; Herbert 2005). Moreover, Di Bella et al., (2006) have noted that several genes putatively implicated in the etiology of autism such as CDC16 (Di Bella et al ., 2006), NF1 (Gillberg and Forsell, 1984), H-RAS (Herault et al., 1995), GRPR (Ishikawa-Brush et al., 1997), TSC1/TSC2 (Crino and Henske, 1999), RAY1/ST7 (Vincent et al., 2002) are also involved in the control of cell cycle and cell growth. Recently, two other genes with similar functions have been added to the list: c-MET (Campbell et al., 2006) and APC (Zhou et al., 2007). Overall, these observations suggest a role of several processes in the etiology of either the fully blown disease or just a sub-clinical expression of certain autistic-like traits or endophenotypes in non autistic subjects.

In summary, whatever is the phenotypic level or developmental stage at which a pathogenetic mechanism is considered, a single gene is likely to be less important to the dysfunction than is the whole network (or networks) of genes of which that gene is part. This interpretation has relevant implications for research because it underscores the importance of studying gene networks potentially involved in autism pathogenesis towards the identification of a common cause in a significant number of patients. The very importance of these goals has clearly emerged in the most mature phase of genetic research on autism. The actual realization of these goals will probably rely on the new approaches that can now be exploited on this disorder thanks to the advent of genomic research.

What Lessons Did We Learn from Genetic Studies on Autism ?

There are several excellent reviews on the genetics of autism (Yang and Gill, 2007; Happè et al., 2006; Veenestra-VanderWeele and Cook, 2004; Muhle et al., 2004; Folstein and Rosen-Sheidley, 2001) thus making dispensable the need for us to describe and discuss in detail the results of these studies. Overall, molecular genetics and cytogenetic studies performed so far, strongly support the notion that genetically autism is a very heterogenous disorder. Genetic heterogeneity for such a complex disease would imply that it is unlikely that any two patients would share the same susceptibility genes. This is the main reason why it has been so difficult or impossible to replicate linkage and association studies.

In our opinion, the best available data generated by a genetic research in these years have concerned the studies on (i) chromosomal rearrangements, (ii) monogenic diseases associated to autism and (iii) few association studies. Integrating the information contributed by these studies has been instrumental to propose some models of autism pathogenesis. For example, the data on neuroligins nos. 3 and 4, neurexin, FMR1, GluR6, MecP2 and SHANK3 genes have been integrated in a general pathogenetic model of autism based on alterations of synaptogenesis and synaptic function, specifically of the glutamatergic synapse (Zoghbi, 2003; Persico and Bourgeron, 2006; Belmonte and Bourgeron, 2006; The Autism Genome Project Consortium, 2007).

How many autism genes are there ? In the whole population of ASD patients this number is probably very high, perhaps hundreds (Pritchard, 2001). In contrast, the number of susceptibility genes in one patient would be 2-10 (Pickles et al., 1995) but as stated by Folstein and Rosen-Sheidley (2001) these genes would rather differ among patients. This latter hypothesis is consistent with the observations done by cytogenetic and molecular genetics studies.

The Genomic Bases of Autism

Altogether, the above considerations underscore the need for novel and more powerful approaches to address the conceptual and technical challenges posed by the biological and genetic research on this complex disorder.

Genomic research offers an unprecedented opportunity to study autism deep into its cellular, sub-cellular and molecular roots. These perspectives have been opened by the development of microarrays and bioinformatics just to mention few relevant technological advances. There are many ways of assessing structural and functional genomic changes by microarray technology. For the scope of this article, however, we will only consider Copy Number Variations and modulated gene expression profiles because these are the type of changes that have been mostly studied in autistic patients in recent years.

Copy Number Variations (CNVs)

CNVs consist in the loss or gain of specific DNA segments containing a variable number of genes. This number depends on the gene density and size of the chromosomal region that has been duplicated or deleted. These cryptic chromosomal rearrangements may result from a *de novo* mutation, occurred in one of the two parents' germ cells during gametogenesis. Alternatively, these rearrangements can be transmitted from one or both "unaffected" parents to the affected siblings. Individuals can be homozygous or heterozygous for a specific CNV.

It is not yet clear if CNVs detected in unaffected individuals have to be considered unexpressed phenotypically. On one hand, it may be anticipated that not all CNVs detected in a patient are etiologically relevant for the specific phenotype under study (in our case autism). On the other hand, gene dosage for one or more genes may affect specific cellular, sub-cellular or molecular functions with only a subclinical manifestation. Furthermore, CNV mutations may either act throughout the entire life of the patient, from embryonic to adult life, or at specific critical developmental windows. A potentially complicating factor when interpreting the possible phenotypic effects of a CNV mutation depends on the different roles that a gene may play at different developmental stages and in different tissues. Finally, with currently used microarray technology it will be very difficult to detect all CNVs present in the genome of a patient. In addition, CNVs probably represent only a fraction of all disease-causing mutations borne by the patient. Pathogenetic point mutations such as those affecting the coding sequence might indeed co-exist with CNVs in the genome of a patient.

To date, several studies have reported the identification of many Copy Number Variations in the gene-rich regions of the genome in autistic patients (The Autism Chromosome Database: http://projects.tcag.ca/autism/project.html). Recently, at least four microarray studies have reported the identification of many CNVs in large cohorts of autistic patients (Christian et al., 2008; Marshall et al., 2008; The Autism Genome Project Consortium, 2007; Sebat et al., 2007).

Gene Expression Profiling

This type of analysis allows the simultaneous measurements of the levels of mRNA transcribed from all genes active in a given cell type or tissue (transcriptome). However, simply correlating the observed transcriptional changes to the pathogenesis of autism is not straightforward and few relevant conceptual and technical points need to be discussed to clarify the nature and action of several potentially complicating factors. First, the changes we see in the transcriptome of autistic patients' cells are many steps removed from the root causes, the latter ones presumably occurred in early life, i.e., during the CNS development. The effects of those root causes could still be ongoing in the analyzed cells/tissue of a patient and therefore would be potentially detectable. Second, many of the changes in the transcription profiles between cases and controls may not be related at all to the disease, thus increasing the noise in microarray data. These latter changes may result from the action of a variety of factors (genetic, environmental, epigenetic, metabolic etc.). Third, since only changes in the levels of RNAs are monitored, other potentially important pathogenetic changes involving, for example, translational or post-translational modification of proteins (e.g. a missense mutation affecting the catalytic activity or stability of an enzyme) may not be reflected in transcriptional changes and would then go undetected. All these effects would translate in a significant underestimation of the number of genes actually involved in the pathogenesis of autism. Fourth, genetic heterogeneity which is a remarkable feature of autistic disorder in theory implies that the number and type of genes underlying the *same* pathogenetic cellular process (e.g. a signal transduction pathway) would differ among different patients. In addition, an important limitation in the analysis of gene expression profiles of autistic patients consists in the difficulty of carrying this analysis in the brain tissue of living or even deceased patients. Purcell et al., (2001) have carried out a study of gene expression profiles in post-mortem brain tissue from several autistic patients and found specific abnormalities in the levels of mRNAs encoding AMPA-type glutamate receptors and glutamate transporters in the cerebellum. In another study, Samaco et al., (2004) found that MeCP2 mRNA expression is frequently reduced in the (post-mortem) frontal cortex of autistic patients.

An alternative to post-mortem brain tissue has been the use of peripheral blood cells (PBCs), the latter ones easier to obtain from patients. This approach can generate some important information too, assuming that PBCs mirror the transcriptome of neural cells in autistic patients. Baron et al., (2006) performed statistical analysis of gene expression profiles of lymphoblastoid cell lines derived from children with autism and have identified differentially expressed genes between cell lines derived from children with autism and those derived from their normally developing siblings. These genes were then used to identify biochemical pathways potentially involved in autism. Hu et al., (2006) used lymphoblastoid cell lines from monozygotic twins discordant with respect to severity of autism and/or language impairment and found that they exhibit differential gene expression patterns on DNA microarrays. Furthermore, they showed that genes important to the development, structure, and/or function of the nervous system are among the most differentially expressed genes. Nishimura et al., (2007) used microarray analysis to compare the mRNA expression profile in lymphoblastoid cells from males with autism due to a fragile X mutation (FMR1 gene), or a 15q11-q13 duplication (dup 15q), and non-autistic controls. These Authors found that gene expression profiles clearly distinguished autism from controls and separated individuals with autism based on their genetic etiology. They also identified sixty-eight genes that were dysregulated in common between autism with FMR1 mutations and dup(15q).

Other expression profiling data obtained with microarray technology have been deposited in public databases (such as Gene Expression Omnibus database GEO http://www.ncbi.nlm.nih.gov/geo/). With more research groups using the microarray technology more structural and functional genomics data will become available for autistic disorder in the near future.

Interpreting Microarray Data

Once microarray data have been analyzed statistically their interpretation constitutes the main bottleneck towards the identification of biologically meaningful results. Meta-analysis methods have been devised that can help interpreting the data in the context of other gene expression data sets. Such an evaluation may rely on prior biological knowledge specific to the genes that have been picked up by microarray analysis. The biological information on these genes can be extracted using, for example, text mining tools for biological and medical literature (as it has been done for autism by Baron et al., 2006 and Hu et al., 2006) or controlled vocabularies. In the former case the same function may be described using numerous alternative wordings, whereas in the latter this is eliminated.

Gene Ontology

The most widely used controlled vocabulary is Gene Ontology (GO: Ashburner et al., 2000; http://www.geneontology.org/) which describes genes and gene product attributes in any organism, including humans. The three organizing principles of GO are Cellular Component (CC), Biological Process (BP) and Molecular Function (MF). A gene product might be associated with or located in one or more cellular components; it is active in one or more biological processes, during which it performs one or more molecular functions. The terms in an ontology are linked by two relationships: *is_a* and *part_of*. The ontologies are structured as directed acyclic graphs, which are similar to hierarchies but differ from them in that a child, or more specialized term, can have many parents, or less specialized terms.

The frequency of the ontology term within a set of query genes (e.g. the list of genes resulted positive to a CNV array analysis or the list of genes whose expression resulted modulated in a microarray experiment), is compared against the frequency of the same term in a background gene set (e.g. the entire genome or the whole set of genes in the array, respectively). The resulting *p*-value, whose statistical significance is typically evaluated using standard statistical methods, based on the hypergeometric distribution or binomial distribution, illustrates the chance of enrichment of a GO term under the null hypothesis of random selection of the list of genes.

To date the use of Gene Ontology in autism's research has been very limited for the obvious reason that large sets of data, such as those generated by microarray analyses, have not been available. In the study by Yonan et al., 2003, 383 positional candidate genes were selected by genomewide genetic linkage analysis of a large set of families, each with two or more members diagnosed with autism, or autism spectrum disorder (ASD). Using this list, the Authors screened genes for neural-related terms in the Gene Ontology database. By this and other bioinformatics methods they were then able to predict some pathways of interacting genes.

A Gene Ontology Study on ASD

Methods

The main aim of this study was to test the power of GO analysis to help reconstructing biological processes potentially involved in the pathogenesis of autism using microarray data on genes involved in CNVs detected in a population of autistic patients. To this end, we have performed a meta-analysis by which data from each patient were combined to generate a list of genes that was then used to query the GO database. We present here the preliminary results of this study.

CNV Data

We have compiled a list of genes comprised in the intervals defined by the CNVs detected in the autistic population studied by The Autism Genome Project Consortium, (2007). The latter Authors analyzed the DNA of each individual from a large cohort of autistic patients to look for the occurrence of such rearrangements using the Affymetrix GeneChip Mapping 10K 2.0 microarray. The information about number, size, start and end points of each CNV is contained in the Supplementary table 4 of The Autism Genome Project Consortium, (2007) and has been downloaded from he Nature website. We have used the data referring to 35 patients whom all CNVs have been validated and subjected to filtered analysis. The genes included in each CNV were identified and retrieved manually using the information contained in the NCBI Mapeviewer (Build 35.1: http://www.ncbi.nlm.nih.gov/). 1750 genes were identified. The approach we have followed to search the GO database is analogous to the Over Representation Analysis (ORA) (Pavlidis et al., 2004; Draghici et al., 2003). Using the hypergeometric distribution a *p*-value was associated to each GO term. This allows to select the ontology terms that do not comply with the null hypothesis of a randomly selected list of genes. Moreover, the *p*-values were also used to score each GO term (Draghici et al., 2003). The genes contained in the HUGO database (http://www.gene.ucl.ac.uk/ nomenclature) were used as a reference set of genes. The selected GO terms are those resulting not consistent with the random null hypothesis at a confidence level of 5% with *p*-values corrected for multiple comparisons (Draghici, 2003).

Results

GO analysis performed on genes enriched within CNVs, uncovered many statistically significant GO terms in the BP, MF and CC categories. In particular, we have selected 22 of such terms in the BP category that are related to neurophysiological processes and central nervous system development (shown in the table).

We have also asked how many patients are contributing information to the identification of a specific GO term. For the GO terms we have analyzed so far, the number of patients involved appears to be very variable and rarely involves one patient only (data not shown). This latter case applies to large dels/dups causing loss or gain of a high number of genes.

Conclusion

With only few exceptions, the many genetic studies performed on autism in past years have not led to a clear understanding of autism pathogenesis. Genetic heterogeneity has been considered by many authors a major factor accounting for this poor outcome, especially for those studies that have used linkage analysis. Other theoretical considerations also suggest that the one-by-one search of candidate genes in autism may not be a conceptually appropriate approach to cope with the many challenges posed by the complex biological structure of this disorder. In this problematic context, some Authors have offered new and interesting perspectives on autism pathogenesis. We present below excerpts from two recent inspiring reviews:

> "….The approach in most studies has been to attempt to dissect autism as if it were a lesion, a missing locus or capacity within an otherwise normal, fully developed brain in which all other factors have somehow been held constant. This approach is inappropriate to the study of developmental disorders because it assumes that the disorder is a function of a localised module, rather than an emergent property of developmental interactions between many brain regions and functions…" (Belmonte et al., 2004b).

> "….In networks of interacting neurons, just as in networks of interacting genes, dosage effects are crucial, and more is not always better… All of these genes participate in setting a level of bias within a continuum of neural connectivity, establishing a sort of 'neural dosage' analogous to gene dosage. Optimizing this neural dosage is crucial in optimizing the representational capacity of neural networks…Thus either extreme in the dosage of a neuronal property may produce the same neural abnormality. This realization may be crucial to making sense of autism's multifactorial genetics and divergent endophenotypes…Considering autism at the network level may lead to a useful dissociation of therapy from pathology, in that normalizing influences may be applied to the network via mechanisms entirely distinct from those that have disrupted it…" (Belmonte and Bourgeron, 2006).

How would one translate these novel and attractive views and interpretations on autism pathogenesis to scientific research, especially to genetic research ? We suspect that such a translation will probably require a theorethical as well as a methodological paradigm shift by which the strategy based on the search for candidate genes will be gradually replaced by the search for candidate biological processes or pathways. Genomic research with its inventory of technological and analytical tools represents one of the best approaches available to date to assist this transition. Based on these considerations, we have recently started structural and functional genomics studies in ASD to gain new information and insights on the molecular bases of this disorder. In this chapter we are reporting the preliminary results of a comprehensive Gene Ontology analysis of the genes involved in CNVs detected in autistic patients. To our knowledge this is the first extended GO analysis performed for ASD.

Our meta-analysis identified a wide range of functional categories related to various multicellular, cellular, sub-cellular and molecular processes. Interestingly, a portion of these GO terms are involved in neurophysiological and cellular processes critical for the development of the CNS. Many phases of the development are represented, from early neurogenesis/gliogenesis to myelination (see the table). How these findings could be interpreted and used ? An important point to bear in mind (discussed already in the Introduction), is that changes in dosage for certain genes may not be expressed phenotypically or simply are not related at all to autism. Thus, the results of a GO analysis will only provide information on the "candidate", rather than the "actual", biological processes involved in the pathogenesis of autism. The candidate biological process could then be confirmed by analyzing different sets of microarray data (i.e., CNV or expression profiling) obtained from distinct autistic populations.

The identification of many functional categories could be interpreted as the effect of the marked heterogeneity characterizing the autistic population we have analyzed. This hypothesis is consistent with the fact that the population studied by The Autism Genome Project Consortium (2007) was not stratified according to a specific endophenotype. In future studies GO could be used to analyze microarray data from patients stratified according to the presence/absence of certain endophenotypes to test, for example, if CNVs detected in epileptic autistic patients are enriched in genes involved in the synthesis and metabolism of excitatory and/or inhibitory neurotransmitters and their receptors.

The actual involvement of the genes in a network, underlying a specific function mapped to GO, could be verified experimentally by decreasing or increasing (depending on the specific question asked) their expression in suitable model systems. For example, RNA interference of genes in a candidate network could be used in cultured neuronal cells to monitor possible morphological and functional changes.

Table. A selection of GO terms identified by genes involved in Copy Number Variations detected in autistic patients (data from The Autism Genome Project Consortium, 2007)

GO code	GO classes	*p*-value*
GO:0007399	nervous system development	0.006
GO:0007417	central nervous system development	0.007
GO:0022008	Neurogenesis	0.006
GO:0048699	generation of neurons	0.006
GO:0014017	neuroblast fate commitment	0.004
GO:0007400	neuroblast fate determination	0.004
GO:0014016	neuroblast differentiation	0.004
GO:0042063	Gliogenesis	0.006
GO:0014014	negative regulation of gliogenesis	0.007
GO:0043350	neuroblast proliferation (sensu Vertebrata)	0.006
GO:0008347	glial cell migration	0.003
GO:0008045	motor axon guidance	0.007
GO:0030182	neuron differentiation	0.007
GO:0016358	dendrite development	0.004
GO:0019226	transmission of nerve impulse	0.007
GO:0019228	generation of action potential	0.003
GO:0007268	synaptic transmission	0.007
GO:0016189	synaptic vesicle to endosome fusion	0.004
GO:0035249	synaptic transmission, glutamatergic	0.002
GO:0009450	gamma-aminobutyric acid catabolic process	0.007
GO:0009448	gamma-aminobutyric acid metabolic process	0.007
GO:0031641	regulation of myelination	0.007

**p*-values have been corrected for multiple comparisons.

In the autistic population we have analyzed with GO, very few patients share the same CNVs, and therefore the same genes. This suggests that the "common denominator" of autism pathogenesis, among different patients, is unlikely to reside on a common gene or even a set of genes. Autistic patients would rather share the same GO term (or biological process). We are currently testing this model using expression profiling data from blood cells of autistic patients to assess the degree of gene-sharing for the same GO term in different patients.

In conclusion, it can be anticipated that, in the near future, biological research on autism will strongly rely on genomic studies. Within this research framework, Gene Ontology is especially suited as a "hypotheses-building" tool that can inform research on this disorder. Finally, we suggest that functional categories (or ontology terms), rather than individual genes, are more suitable and useful "units" to describe and dissect experimentally the biological structure of autism.

References

Ashburner, M; Ball, CA; Blake, JA (2000). Gene Ontology: tool for the unification of biology. The Gene Ontology consortium, *Nature Genetics,* 25, 25-29.

Baron, CA; Liu, SY; Hicks C; Gregg, JP (2006). Utilization of lymphoblastoid cell lines as a system for the molecular modeling of autism. *J. Autism Dev. Disord.* 36, 973-82.

Belmonte, MK; Allen, G; Beckel-Mitchener, A; Boulanger, LM; Carper, RA; Webb, SJ (2004a). Autism and Abnormal Development of Brain Connectivity. *The Journal of Neuroscience.* 24, 9228-31.

Belmonte, MK; Cook, EH; Anderson, GM; Rubenstein, JL; Greenough, WT; Beckel-Mitchener, A; Courchesne, E; Boulanger, LM; Powell, SB; Levitt, PR; Perry, EK; Jiang, YH; DeLorey, TM; Tierney, E (2004b). Autism as a disorder of neural information processing: directions for research and targets for therapy. *Mol. Psychiatry.* 9, 646-63.

Belmonte, MK & Bourgeron, T (2006). Fragile X syndrome and autism at the intersection of genetic and neural networks. *Nat. Neurosci.* 9, 1221–25.

Bird, CM; Castelli, F; Malik, O; Frith, U; Husain, M (2004). The impact of extensive medial frontal lobe damage on 'Theory of Mind' and cognition. *Brain.* 127, 914-28.

Campbell, DB; Sutcliffe, JS; Ebert, PJ; Militerni, R; Bravaccio, C; Trillo, S; Elia, M; Schneider, C; Melmed, R; Sacco, R; Persico, AM; Levitt, P (2006). A genetic variant that disrupts MET transcription is associated with autism. *Proc. Natl. Acad. Sci. U S A.* 45, 16621-32.

Christian, SL; Brune, CW; Sudi, J; Kumar, RA; Liu, S; Karamohamed, S; Badner, JA; Matsui, S; Conroy, J; McQuaid, D; Gergel, J; Hatchwell, E; Gilliam, TC; Gershon, ES; Nowak, NJ; Dobyns, WB; Cook, EH Jr.(2008) Novel sub-microscopic chromosomal abnormalities detected in autism spectrum disorder. *Biol Psychiatry*. 63, 1111-7.

Crino, PB & Henske, EP (1999). New developments in the neurobiology of the tuberous sclerosis complex. *Neurology*. 53, 1384–90.

Di Bella, MA; Cali, F; Seidita, G; Mirisola, M; Ragusa, A; Ragalmuto, A; Galesi, O; Elia, M; Greco, D; Zingale, M; Gambino, G; D'Anna, RP; Regan, R; Carbone, MC; Gallo, A; Romano, V (2006). Screening of subtelomeric rearrangements in autistic disorder: identification of a partial trisomy of 13q34 in a patient bearing a 13q;21p translocation. *Am. J. Med. Genet. Part B Neuropsychiatr. Genet.* 141, 584-90.

Draghici, S. Data Analysis Tools for DNA Microarrays. Chapman & Hall/CRC; Har/Cdr Re edition; June 4, 2003.

Draghici, S; Khatri, P; Martins, RP; Ostermeier, GC; Krawetz, SA (2003). Global functional profiling of gene expression. *Genomics.* 81, 98-104.

Flint, J & Munafo, MR (2007). The endophenotype concept in psychiatric genetics. *Psychol. Med.* 37, 163-80.

Folstein SE & Rosen-Sheidley B (2001). Genetics of autism: Complex aetiology for a heterogeneous disorder. *Nat. Rev. Genet.* 9, 943–55.

Frith, U. Autism: explaning the enigma. 2nd edition. Oxford. Blackwell Publishing; 2003.

Gillberg, C & Forsell, C (1984). Childhood psychosis and neurofibromatosis: more than a coincidence? *J. Autism. Dev. Disord.* 14, 1–8.

Happé, F; Ronald, A; Plomin, R (2006). Time to give up on a single explanation for autism. *Nat. Neurosci.* 9, 1218-20.

Herault, J; Petit, E; Martineau, J; Perrot, A; Lenoir, P; Cherpi, C; Barthelemy, C; Sauvage, D; Mallet, J; Muh, JP (1995). Autism and genetics: clinical approach and association study with two markers of HRAS gene. *Am. J. Med. Genet.* 60, 276-81.

Herbert, MR (2005). Large Brains in Autism: The Challenge of Pervasive Abnormality. *The Neuroscientist.* 5, 417-40.

Hu, WH; Bryan, CF; Heine, S; Lee, NH; Quackenbush, J (2006). Gene expression profiling of lymphoblastoid cell lines from monozygotic twins discordant in severity of autism reveals differential regulation of neurologically relevant genes. *BMC Genomics.* 18, 7-18.

Ishikawa-Brush, Y; Powell, JF; Bolton, P; Miller, AP; Francis, F; Willard, HF; Lehrach, H; Monaco, AP (1997). Autism and multiple exostoses associated with an X;8 translocation occurring within the GRPR gene and 30 to the SDC2 gene. *Hum. Mol. Genet.* 6, 1241-50.

Jamain, S; Quach, H; Betancur, C; Rastam, M; Colineaux, C; Gillberg, IC; Soderstrom, H; Giros, B; Leboyer, M; Gillberg, C; Bourgeron, T; Paris. Autism Research International Sibpair Study. (2003). Mutations of the X-linked genes encoding neuroligins NLGN3 and NLGN4 are associated with autism. *Nat. Genet.* 34, 27-9.

Koshino, H; Kana, RK; Keller, TA; Cherkassky, VL; Minshew, NJ; Just, MA (2007). fMRI Investigation of Working Memory for Faces in Autism: Visual Coding and Underconnectivity with Frontal Areas. *Cereb. Cortex.* May 20; [Epub ahead of print]

Marshall, CR; Noor, A; Vincent, JB; Lionel, AC; Feuk, L; Skaug, J; Shago, M; Moessner, R; Pinto, D; Ren, Y; Thiruvahindrapduram, B; Fiebig, A; Schreiber, S; Friedman, J; Ketelaars, CE; Vos, YJ; Ficicioglu, C; Kirkpatrick, S; Nicolson, R; Sloman, L; Summers, A;, Gibbons, CA; Teebi, A; Chitayat, D; Weksberg, R; Thompson, A; Vardy, C; Crosbie, V; Luscombe, S; Baatjes, R; Zwaigenbaum, L; Roberts, W; Fernandez, B; Szatmari, P; Scherer, SW. (2008). Structural variation of chromosomes in autism spectrum disorder. *Am J Hum Genet.* 82, 477-88.

Muhle, R; Trentacoste, SV; Rapin, I (2004). The genetics of autism. *Pediatrics.* 113, 472-86.

Nishimura, Y; Martin, CL; Lopez, AV; Spence, SJ; Alvarez-Retuerto, AI; Sigman, M; Steindler, C; Pellegrini, S; Schanen, NC; Warren, ST; Geschwind, DH (2007). Genome wide expression profiling of lymphoblastoid cell lines distinguishes different forms of autism and reveals shared pathways. *Hum. Mol. Genet.* May 21; [Epub ahead of print].

Pavlidis, P; Qin, J; Arango, V; Mann, JJ; Sibille, E (2004). Using the Gene Ontology for Microarray Data Mining: A Comparison of Methods and Application to Age Effects, in Human Prefrontal Cortex. *Neurochemical Research.* 29, 1213-22.

Persico, AM & Bourgeron, T (2006). Searching for ways out of the autism maze: genetic, epigenetic and environmental clues. *TRENDS in Neurosciences.* 29, 349-58.

Pickles, A; Bolton, P; Macdonald, H; Bailey, A; Le Couteur, A; Sim, CH; Rutter, M (1995). Latent class analysis of recurrence risks for complex phenotypes with selection and measurement error: a twin and family history study of autism. *Am. J. Hum. Genet.* 57, 717–26.

Pritchard, JK (2001). Are rare variants responsible for susceptibility to complex diseases? *Am. J. Hum. Genet.* 69,124–37.

Purcell, AE; Jeon, OH; Zimmerman, AW; Blue, ME; Pevsner, J (2001). Postmortem brain abnormalities of the glutamate neurotransmitter system in autism. *Neurology.* 57, 1618-28.

Samaco, RC; Nagarajan, RP; Braunschweig, D; LaSalle JM (2004). Multiple pathways regulate MeCP2 expression in normal brain development and exhibit defects in autism-spectrum disorders. *Hum. Mol. Genet.* 13, 629-39.

Saskia, J; Palmen, MC; Van Engeland, H; Hof, PR; Schmitz, C (2004). Neuropathological findings in autism. *Brain.* 127, 2572–83.

Sebat, J; Lakshmi, B; Malhotra, D; Troge, J; Lese-Martin, C; Walsh, T; Yamrom, B; Yoon, S; Krasnitz, A; Kendall, J; Leotta, A; Pai, D; Zhang, R; Lee, YH; Hicks, J; Spence, SJ; Lee, AT; Puura, K; Lehtimaki, T; Ledbetter, D; Gregersen, PK; Bregman, J; Sutcliffe, JS; Jobanputra, V; Chung, W; Warburton, D; King, MC; Skuse, D; Geschwind, DH; Gilliam, TC; Ye, K; Wigler, M (2007). Strong association of de novo copy number mutations with autism. *Science.* 316, 445-49.

The Autism Genome Project Consortium (2007). Mapping autism risk loci using genetic linkage and chromosomal rearrangements. *Nat. Neurosci.* 39, 319-28.

Veenestra-VanderWeele J & Cook, EH Jr (2004). Molecular genetics of autism spectrum disorders. *Mol. Psychiatry.* 9, 819-32.

Vincent, JB; Petek, E; Thevarkunnel, S; Kolozsvari, D; Cheung, J; Patel, M; Scherer, SW (2002). The RAY1/ST7 tumor-suppressor locus on chromosome 7q31 represents a complex multi-transcript. *Genomics.* 80, 283–94.

Yang, MS & Gill, M (2007). A review of gene linkage, association and expression studies in autism and an assessment of convergent evidence. *Int. J. Dev. Neurosci.* 25, 69-85.

Yonan, AL; Palmer, AA; K. C. Smith, KC; Feldman, I; Lee, HK; Yonan, JM; Fischer, SG; Pavlidis, P; Gilliam, TC (2003). Bioinformatic analysis of autism positional candidate genes using biological databases and computational gene network prediction. *Genes, Brain and Behavior.* 2, 303–20.

Zhou, XL; Giacobini, M; Anderlid, BM; Anckarsater, H; Omrani, D; Gillberg, C; Nordenskjold, M; Lindblom, A (2007). Association of adenomatous polyposis coli (APC) gene polymorphisms with autism spectrum disorder (ASD). *Am. J. Med. Genet. Part B Neuropsychiatr. Genet.* 144, 351-54.

Zoghbi, HY (2003). Postnatal neurodevelopmental disorders: meeting at the synapse? *Science.* 302, 826-30.

In: Causes and Risks for Autism
Editors: A. C. Giordano et al.
ISBN: 978-1-60456-861-5

Chapter III

Gastropods as an Animal Model for Studying Autism and Other Behavioral Phenotypes

Jeff S. Hatfield
USGS Patuxent Wildlife Research Center
12100 Beech Forest Road, Laurel, Maryland 20708

Abstract

Gastropods have an unusual genetic mechanism in which the mother's genotype determines the shell phenotype of her offspring. Experiments have shown that these unknown, maternal-effect genes determine shell chirality, the direction each snail shell coils, with dextral or right handed having a clockwise spiral, and sinistral or left handed having a counterclockwise spiral. Most snail species consist only of dextral individuals, while some species are entirely sinistral, and some species are polymorphic for shell chirality, producing both dextrals and sinistrals in the same species. This particular genetic mechanism implies a lack of homozygous-dominant (*R/R*) individuals in the alternative shell phenotype, while the primary phenotype can have any of the genotypes of homozygous dominant, heterozygous (*R/r*), or homozygous recessive (*r/r*) in snails. In humans, the hair-whorl rotation, the direction in which the hair spins at the back of the head, is generally either clockwise or counterclockwise and has been associated with handedness, cerebral laterality, and sexual orientation. As such, direction of hair-whorl rotation is thought to be a phenotypic marker underlying the genetics of various behavioral phenotypes in our species. Previous research has already associated handedness with sexual orientation, psychosis, and autism spectrum disorders (ASD), but to the best of my knowledge, hair-whorl rotation has not been tabulated in autistic individuals. A recent theory has proposed that maternal-effect genes are involved in determining handedness and hair-whorl rotation in humans, in addition to various behavioral phenotypes in our species. These genes (possibly *RHD* and *RHCE*) predict a lack of homozygous-dominant individuals among the alternative phenotypes in humans, analogous to shell chirality in snails. If maternal-effect genes are interacting with biparentally-expressed genes to cause some cases of ASD and other behavioral phenotypes in humans, then current genomic search results (i.e., linkage mapping using single nucleotide polymorphisms) for all of these genes will be obscured. To increase precision of genome searches during sibling studies, researchers should search for maternal-effect genes (i.e., never or almost never homozygous in the alternative phenotype; alleles shared *less* commonly than expected by chance in sets of affected siblings) in addition to biparentally-expressed genes (i.e., genes always or almost always homozygous in the alternative phenotype; alleles shared *more* commonly than expected by chance in siblings). If it can be shown that the snail chirality gene is homologous to the human handedness and hair-whorl genes, even if these genes turn out not to be *RH* genes, snails may still provide an animal model for evaluating possible environmental causes of ASD in humans. For example, aquatic snails, many of which have a short generation time and breed easily in captivity, could be treated with mercury to test the expression of the opposite chirality in their offspring. Other chemical agents or mixtures could also be tested in snails to determine possible correlates to environmental exposure in humans.

Introduction

A maternal-effect gene is defined as one in which the mother's genotype determines the phenotype displayed by her offspring. The best known and most studied example of such genes comes from Gastropods, or snails, but similar genes are also known to exist in other species (Beeman and Friesen 1999). Snails are hermaphroditic, so each snail possesses both male and female sex organs and thus each individual can serve either as mother (the individual that produces the eggs) or father (the individual that fertilizes the eggs). Snail shells are either right handed (dextral, clockwise spiral from apex of shell) or left handed (sinistral, counterclockwise spiral from apex of shell) in their development, and it is the unknown gene that determines handedness or chirality in snails that functions as a maternal-effect gene (Schilthuizen and Davison 2005). Most snail species consist only of dextral individuals, but some species are entirely sinistral, and some species are polymorphic for shell chirality, producing both dextrals and sinistrals within the same species (analogous to humans producing individuals who are either right handed or non-right handed).

The manner in which this genetic mechanism functions in snails is that the mother snail deposits a substance into each of her eggs that determines the chirality of the snails that develop from those eggs. This substance has never been identified, but it is known to exist because of the results from experimental studies. This unknown substance's composition depends on the mother's genes, however, not the offspring's, yet it somehow guides the development of asymmetry (and ultimately, chirality) in her offspring. If a snail species is all or mostly dextral, then dextral is defined as the primary phenotype for that species. If some individuals of that species are also sinistral, then this is defined as the alternative phenotype for that species. Similarly, in humans, the primary phenotype is defined as right handed because this phenotype is more common and the alternative phenotype is then defined as non-right handed. One can also extend the definition of primary phenotype in humans by defining the primary phenotype as right handed, left brained (i.e., cerebral laterality, the primary phenotype has left-sided language lateralization), clockwise hair whorl, heterosexual, non-psychotic, non-speech-dyslexic, and non-autistic. Thus, the primary phenotype is more common among individuals of our species and the alternative phenotypes are much less so. These phenotypes will be discussed in more detail below, under the human model.

If a dextral snail is either homozygous dominant (*R/R*) or heterozygous (*R/r*) for the chirality gene, as discussed above, she produces and deposits a substance into her eggs that causes them to develop into individuals whose shells spiral dextrally. If she is homozygous recessive (*r/r*), however, then her eggs will become all or mostly sinistral individuals (Freeman and Lundehus 1982). Thus it is the mother's genes that determine the direction in which her offspring's shells coil, not their own genes. The reason it is important for snails to regulate the chirality of their shells is because many species, particularly those with a high spire, have difficulty mating with individuals of the opposite chirality (Schilthuizen and Davison 2005). Presumably some population genetic reason, such as greater fitness in individuals with the specified chirality, explains why the mother snail is causing her offspring to have similar chirality among themselves via her unknown, maternal-effect genes.

One prediction from this model is that the alternative phenotype will never be homozygous dominant. The primary phenotype can be homozygous dominant, heterozygous, or homozygous recessive, but the alternative phenotype can only be heterozygous or homozygous recessive. Thus, the alternative phenotype is never homozygous dominant, at least in the genetic etiologies, and this is a prediction that can be used in searching or testing for potential candidate genes.

Human Model

Recently, I proposed a similar maternal-effect genetic model for handedness in humans, with some important differences, as discussed in Hatfield (2006). Regardless of the differences, this model leads to the same prediction in humans as in snails (i.e., a lack of homozygous-dominant individuals among the alternative phenotypes). It should be noted that any such gene that operates this way would probably lead to this same prediction. However, it is an unusual and, therefore, interesting prediction. It is also testable.

Klar (2003) proposed a model for the gene that determines handedness and hair-whorl rotation in humans. According to this model, humans are either right handed or non-right handed, while hair-whorl rotation (i.e., the direction the scalp hair turns at the back of the head) can be either clockwise, counterclockwise, or in rare cases, even more complicated. Most people (> 95%) have a single, dominant hair whorl, either clockwise or counterclockwise, but some people have double hair whorls, with both clockwise and counterclockwise hair whorls on the same scalp (Wunderlich and Heerema 1975), and other rare patterns also exist (R. Lippa, personal communication). Klar (2003) proposed that a common genetic mechanism controls both handedness and hair-whorl rotation in humans (with this result being replicated by Beaton and Mellor 2007), that this gene was possibly also related to cerebral laterality (with evidence provided later by Weber et al. 2006), and may also be involved in speech dyslexia (e.g., stuttering) in some individuals. He further proposed that this unknown, *RGHT* gene determines (or at least is somehow involved in determining) the various phenotypes for these traits in our species. Klar (2004a) speculated that this same gene may be implicated in the etiology of some forms of mental illness (specifically, bipolar disorder and schizophrenia) and Klar (2004b) inferred that this putative gene is associated with sexual orientation in humans by demonstrating an empirical association between hair-whorl patterns and male sexual orientation. Considerable earlier research demonstrated a similar association between handedness and sexual orientation, with homosexual individuals displaying higher rates of non-right handedness than heterosexual individuals (Lalumière et al. 2000, Lippa 2003).

It is informative to describe in more detail the mechanism that Klar (2003) has proposed for how the *RGHT* gene leads to asymmetries in the development our species and others. Specifically, his model provides a mechanism for differentiation of particular cell types during mitosis (Klar 2004b, Armakolas and Klar 2006, 2007). Klar believes that the *RGHT* gene acts to cause non-random segregation of DNA strands on chromosome 11 during mitosis in our species, and this leads to differentiation of certain cell lines in the developing embryo, which ultimately leads to asymmetrical placement of brain structures in humans. Because of the non-random segregation of DNA strands during mitosis, some cells receive different genetic information relative to other cells. Thus, this epigenetic mechanism allows for the on-off switches of the genes to be different between the daughter cells due to the non-random segregation of the DNA chains during mitosis, ultimately leading to differentiation into certain cell types. Klar's proposed mechanism is unorthodox and surprising, however, as it contradicts the conventional assumption that DNA strands are allocated randomly to daughter cells during mitosis.

Klar's genetic model operates as follows. If an individual is homozygous recessive (*r/r*) for the *RGHT* gene, he or she develops handedness and hair-whorl rotation at random during fetal development. Thus, 50% of these individuals will be right handed, and 50% will be non-right handed, and independently, 50% will have clockwise hair whorl, and 50% will have counterclockwise hair whorl. Klar termed this process the *random recessive* pattern of brain development. If an individual is homozygous dominant (*R/R*) or heterozygous (*R/r*), then these individuals develop clockwise hair whorl and become right handed according to Klar's model. Klar does not have an explanation for more complicated hair whorls, however, and excluded such individuals from his research.

My theory (Hatfield 2006) is an extension of Klar's model in which I proposed maternal-effect genes (possibly *RHD* and *RHCE*) determine handedness and hair-whorl rotation and are also involved in sexual orientation and other behavioral phenotypes in humans, with maternal immunization being the mechanism for how the maternal effect is realized. Interestingly, *RHD* was recently linked to sexual orientation in humans (Ellis et al. 2008). In snails, the mother determines the chirality of her offspring, so the father's genes are not important for determining chirality in snails. In humans, however, assuming my model is correct, the father's genes do sometimes have an effect, depending on whether the mother maternally immunizes or not. Thus, first-born children, for example, would tend to have their handedness and hair-whorl rotation determined by both the father's and the mother's genes following Klar's model, but as the mother bears more children, and if she maternally immunizes before or during the later pregnancies, then her genes are more important for determining the phenotypes of these later-born children according to my model.

Thus, first-born children may follow Klar's model fairly closely, but later-born children deviate from this model more and more, as maternal immunization becomes stronger. This is the maternal effect. Some mothers may not maternally immunize at all, so these families would also follow Klar's model. My extension of Klar's model basically implies that some heterozygous individuals also go through the random-recessive pattern of choosing handedness and hair-whorl rotation at random, due to maternal immunization by their mothers. More complicated models are also possible.

In Hatfield (2006), I proposed that autism spectrum disorders (ASD) in humans may be associated with our maternally-immunizing genes (specifically *RHD* and *RHCE*, with a related hypothesis being that *RHD* is the *RGHT* gene). Asbury (2006) also proposed that the *RGHT* gene may be associated with ASD. Indeed, handedness has been shown to be associated with ASD (Bryson 1990), as well as with sexual orientation (Lippa 2003) and psychosis (Klar 2003). The fact that more males than females display all of these alternative phenotypes (except bipolar disorder, apparently) provides some evidence for maternal immunization as well, although certainly other explanations are possible, such as the effects of a sex-linked gene and prenatal variations in exposure to sex hormones. Evidence for the maternal-immunization hypothesis comes from studies of the fraternal birth-order effect found in homosexual males (i.e., the consistently demonstrated phenomenon that each additional older brother increases a man's odds of being homosexual; Blanchard 2004, as discussed in Hatfield 2006) and a similar birth order effect found with handedness in chimpanzees (Hopkins et al. 2005, discussed by Wolman 2005). These findings, coupled with evidence from the human monozygotic twin-concordance studies, may help suggest possible biological models.

Of course, a model is of limited value unless data are eventually collected to test the model and estimate its parameters, but existing results suggest *a priori* that certain models may be more appropriate than others, if maternally-immunizing genes are involved as maternal-effect genes in humans. An interaction of three maternal-effect genes (acting the way *RHD* does, in which male fetuses are more likely to initiate maternal immunization than female fetuses; see Blanchard 2004) working as Klar (2003) and Hatfield (2006) proposed, would lead to many more males than females displaying the alternative phenotype (ASD, in this case) and it could lead to a twin-concordance prediction of 7/8 or 87.5%, if all of the seven possible alternative phenotypes are more susceptible to ASD and the primary phenotype (i.e., right handed and left brained) is not susceptible. This value of 87.5% is within the range 60-95% given in the literature recently (Schellenberg et al. 2006) for monozygotic twin concordance for ASD, but certainly other models are possible. Just one maternal-effect gene or two interacting maternal-effect genes can lead to a twin-concordance prediction of 50%, as is found in the literature on sexual orientation (Lalumière et al. 2000), but depending on the model, can also yield twin-concordance predictions of 25% or 75%.

If maternal-effect genes are involved in producing the alternative phenotypes discussed above, then they are likely not the only genes involved. Indeed, even a maternally-suppressed paternally-associated gene (*LRRTM1*) was recently found to be involved in handedness and schizophrenia in humans via an epigenetic mechanism (Francks et al. 2007). Other studies of schizophrenia have implicated a number of contributing genes (e.g., see Law et al. 2006, Barnett et al. 2007, Bellon 2007), in addition to *RHD* (Palmer et al. 2002), and similar studies have been conducted for ASD (e.g., see Ylisaukko-oja et al. 2005, Campbell et al. 2006). In genetic terms, then, there may be several (or many) different etiologies leading to the same phenotype, or alternatively, there may be an interaction among maternal-effect genes and biparentally-expressed genes (i.e., autosomal, non-maternal-effect genes) to cause the development of such phenotypes. It is certainly possible that there are some (or many) environmental causes for the development of these phenotypes as well, as discussed in Hatfield (2006), and it is also possible that genetic factors interact with environmental triggers or factors. Regardless, any compound or chemical agent that interferes with the function of the *RGHT* gene in a developing fetus may be responsible for some cases of ASD.

Genome Searches

One powerful way to search for genes associated with particular traits is to do linkage mapping using single nucleotide polymorphisms (SNPs) from data generated with sibling studies. In such studies, one finds a sample of families that have two or more siblings displaying the phenotype of interest (e.g., ASD, bipolar disorder, homosexuality, schizophrenia, etc.) and search the siblings' genomes for SNPs in which alleles are *more* commonly shared among the siblings than expected by chance in such siblings. Such searches have been fruitful in finding genes in many instances, but for the various alternative phenotypes discussed in this chapter, such searches have not been extremely successful to date (IMGSAC 2001, Levinson et al. 2002, Mustanski et al. 2005, Schellenberg et al. 2006, Suarez et al. 2006, Crow 2007, Baum et al. 2008). These searches look for genes in which the alternative phenotypes are always or almost always homozygous recessive (or dominant). It makes sense to look at this tail of the statistical distribution of such genes (i.e., the right-hand tail), because this is indeed how many biparentally-expressed genes function. However, maternal-effect genes, like those found in snails and proposed in humans, do not fit this model. In fact, it is the opposite tail (ironically, the left-hand tail) of the statistical distribution of such genes that may help locate maternal-effect genes, because this type of maternal-effect gene is never or almost never homozygous dominant in the alternative phenotype. Thus, to find these genes, one will need to search for SNPs in which alleles are *less* commonly shared among the sets of affected siblings than expected by chance in such siblings. Existing SNP data could certainly be re-analyzed to search for such maternal-effect genes.

If the genetic etiologies for these phenotypes include interactions between biparentally-expressed genes and maternal-effect genes (and paternal-effect genes?) these interactions may be difficult to detect with linkage mapping, especially if many different genes or genetic mechanisms are involved. However, it may be possible to detect such a signature with large sample sizes, if one factors into the algorithm that there is an interaction between maternal-effect and biparentally-expressed genes. If environmental pathways are also involved in the development of some or all of the alternative phenotypes, this further complicates the search for relevant genes.

Conclusion

The ideas presented in this chapter and Hatfield (2006) provide the motivation for testing hypotheses concerning whether maternal-effect genes are involved in ASD and other alternative phenotypes in humans. The similarities between the snail chirality gene and the proposed human handedness and hair-whorl genes may just be an analogy, but if true, it is possible that they are homologous as well, not just analogous. Thus, it would be useful to identify the snail chirality gene, as well as the human genes, to determine if they are similar. One hypothesis is that these maternal-effect genes are all *RH* genes (in particular, *RHD* and *RHCE*, in humans). Of course, it is also possible that none of these genes are *RH* genes, but if one finds the snail chirality gene, then one may be able to find the human handedness and hair-whorl genes as well, if they are similar (i.e., homologous) to the snail chirality gene. If all of these genes are eventually demonstrated to be homologous (*RH* genes or otherwise) then snails may provide an animal model for studying the production of the alternative phenotypes in humans (e.g., ASD). Freshwater snails of the genus *Lymnaea* are already used in toxicity studies (Nazrul Islam et al. 2001, Pyatt et al. 2003), and some of these same species of *Lymnaea* are also polymorphic for shell chirality (Schilthuizen and Davison 2005). Retrospective studies of exposure in humans are certainly possible and should be conducted (i.e., estimate exposure of pregnant mothers to some chemical agent of interest and correlate such exposure with, say, ASD rates in various areas). However, even if the retrospective studies are conducted and they suggest a correlation with some chemical agent, one can never do the definitive experimental studies in humans in which one exposes pregnant mothers to hypothesized environmental triggers (mercury, for example) to determine if exposure causes some of their children to display ASD. Thus, it is impossible to prove causation with such retrospective studies, only correlation. However, one can expose snails to mercury to determine if such exposure causes some of their offspring's shells to coil in the opposite direction. Thus, if the snail chirality genes and human handedness and hair-whorl genes are eventually proven to be homologous, snails could be used to test various chemical agents or mixtures to assess their effects on chirality. Snails would then constitute a useful animal model for studying the production of the alternative phenotypes in humans, at least regarding the function (or disrupted function) of the chirality genes.

Acknowledgements

I am grateful to Patrick Connor, U.S. Fish and Wildlife Service, for sending me a paper about snail chirality, and I would also like to thank Michael R. J. Forstner, Texas State University, and Richard A. Lippa, California State University, for many useful discussions concerning the ideas presented in this manuscript. Helpful peer reviews were provided by Natalie Karouna, U.S.

Geological Survey, and Richard Lippa concerning an earlier version of this chapter.

References

Armakolas, A., & Klar, A. J. S. (2006). Cell type regulates selective segregation of mouse chromosome 7 DNA strands during mitosis. *Science, 311,* 1146-1149.

Armakolas, A., & Klar, A. J. S. (2007). Left-right dynein motor implicated in selective chromatid segregation in mouse cells. *Science, 315*, 100-101.

Asbury, C. (2006). Left hand, right brain: The plot thickens. *Cerebrum: The Dana Forum on Brain Science, 7*, 1-14.

Barnett, J. H., Jones, P. B., Robbins, T. W., & Müller, U. (2007). Effects of the catechol-*O*-methyltransferase $Val^{158}Met$ polymorphism on executive function: A meta-analysis of the Wisconsin card sort test in schizophrenia and healthy controls. *Molecular Psychology, 12*, 502-509.

Baum, A. E., Akula, N., Cabanero, M., Cardona, I., Corona, W., Klemens, B., Schulze, T. G., Cichon, S., Rietschel, M., Nöthen, M. M., Georgi, A., Schumacher, J., Schwarz, M., Abou Jamra, R., Höfels, S., Propping, P., Satagopan, J., Detera-Wadleigh, S. D., Hardy, J., & McMahon, F. J., NIMH Genetics Iniative Bipolar Disorder Consortium (2008). A genome-wide association study implicates diacylglycerol kinase eta (DGKH) and several other genes in the etiology of bipolar disorder. *Molecular Psychiatry,13*, 197-207.

Beaton, A. A., & Mellor, G. (2007). Direction of hair whorl and handedness. *Laterality, 12*, 295-301.

Beeman, R. W., & Friesen, K. S. (1999). Properties and natural occurrence of maternal-effect selfish genes ('Medea' factors) in the red flour beetle, *Tribolium castaneum*. *Heredity, 82*, 529-534.

Bellon, A. (2007). New genes associated with schizophrenia in neurite formation: A review of cell culture experiments. *Molecular Psychiatry, 12*, 620-629.

Blanchard, R. (2004). Quantitative and theoretical analyses of the relation between older brothers and homosexuality in men. *Journal of Theoretical Biology, 230*, 173-187.

Bryson, S. E. (1990). Autism and anomalous handedness. In S. Coren (Ed.), *Left handedness: Behavioral implications and anomalies* (pp. 441-456). Amsterdam: Elsevier.

Campbell, D. B., Sutcliffe, J. S., Ebert, P. J., Militerni, R., Bravaccio, C., Trillo, S., Elia, M., Schneider, C., Melmed, R., Sacco, R., Persico, A. M., & Levitt, P. (2006). A genetic variant that disrupts *MET* transcription is associated with autism. *Proceedings of the National Academy of Sciences USA, 103*, 16834-16839.

Crow, T. J. (2007). How and why genetic linkage has not solved the problem of psychosis: Review and hypothesis. *American Journal of Psychiatry, 164*, 13-21.

Ellis, L., Ficek, C., Burke, D., & Das, S. (2008). Eye color, hair color, blood type, and the Rhesus factor: Exploring possible genetic links to sexual orientation. *Archives of Sexual Behavior, 37,* 145-149.

Francks, C., Maegawa, S., Laurén, J., Abrahams, B. S., Velayos-Baeza, A., Medland, S. E., Colella, S., Groszer, M., McAuley, E. Z., Caffrey, T. M., Timmusk, T., Pruunsild, P., Koppel, I., Lind, P. A., Matsumoto-Itaba, N., Nicod, J., Xiong, L., Joober, R., Enard, W., Krinsky, B., Nanba, E., Richardson, A. J., Riley, B. P., Martin, N. G., Strittmatter, S. M., Möller, H.-J., Rujescu, D., St Clair, D., Muglia, P., Roos, J. L., Fisher, S. E., Wade-Martins, R., Rouleau, G. A., Stein, J. F., Karayiorgou, M., Geschwind, D. H., Ragoussis, J., Kendler, K. S., Airaksinen, M. S., Oshimura, M., DeLisi, L. E., & Monaco, A. P. (2007). LRRTM1 on chromosome 2p12 is a maternally suppressed gene that is associated paternally with handedness and schizophrenia. *Molecular Psychiatry, 12,* 1129-1139.

Freeman, G., & Lundehus, J. W. (1982). The developmental genetics of dextrality and sinistrality in the Gastropod *Lymnaea peregra. Wilhelm Roux's Archives, 191*, 69-83.

Hatfield, J. S. (2006). The genetic basis of hair whorl, handedness, and other phenotypes. *Medical Hypotheses, 66,* 708-714.

Hopkins, W. D., Russell, J., Freeman, H., Buehler, N., Reynolds, E., & Schapiro, S. J. (2005). The distribution and development of handedness for manual gestures in captive chimpanzees (*Pan troglodytes*). *Psychological Science, 16*, 487-493.

International Molecular Genetic Study of Autism Consortium (IMGSAC). (2001). A genomewide screen for autism: Strong evidence for linkage to chromosomes 2q, 7q, and 16p. *American Journal of Human Genetics, 69*, 570-581.

Klar, A. J. S. (2003). Human handedness and scalp hair-whorl direction develop from a common genetic mechanism. *Genetics, 165*, 269-276.

Klar, A. J. S. (2004a). A genetic mechanism implicates chromosome 11 in schizophrenia and bipolar disorder. *Genetics, 167*, 1835-1840.

Klar, A. J. S. (2004b). Excess of counterclockwise scalp hair-whorl rotation in homosexual men. *Journal of Genetics, 83*, 251-255.

Lalumière, M. L., Blanchard, R., & Zucker, K. J. (2000). Sexual orientation and handedness in men and women: A meta-analysis. *Psychological Bulletin, 136*, 575-592.

Law, A. J., Lipska, B. K., Shannon Weickert, C., Hyde, T. M., Straub, R. E., Hashimoto, R., Harrison, P. J., Kleinman, J. E., & Weinberger, D. R. (2006). Neuregulin 1 transcripts are differentially expressed in schizophrenia and regulated by 5' SNPs associated with the disease. *Proceedings of the National Academy of Sciences USA, 103*, 6747-6752.

Levinson, D. F., Holmans, P. A., Laurent, C., Riley, B., Pulver, A. E., Gejman, P. V., Schwab, S. G., Williams, N. M., Owen, M. J., Wildenauer, D. B., Sanders, A. R., Nestadt, G., Mowry, B. J., Wormley, B., Bauché, S., Soubigou, S., Ribble, R., Nertney, D. A., Liang, K. Y., Martinolich, L., Maier, W., Norton, N., Williams, H., Albus, M., Carpenter, E. B., deMarchi, N., Ewen-White, K. R., Walsh, D., Jay, M., Deleuze, J.-F., O'Neill, F. A., Papadimitriou, G., Weilbaecher, A., Lerer, B., O'Donovan, M. C., Dikeos, D., Silverman, J. M., Kendler, K. S., Mallet, J., Crowe, R. C., & Walters, M. (2002). No major schizophrenia locus detected on chromosome 1q in a large multicenter sample. *Science, 296*, 739-741.

Lippa, R. A. (2003). Handedness, sexual orientation, and gender-related personality traits in men and women. *Archives of Sexual Behavior, 32*, 103-114.

Mustanski, B. S., DuPree, M. G., Nievergelt, C. M., Bocklandt, S., Schork, N. J., & Hamer, D. H. (2005). A genomewide scan of male sexual orientation. *Human Genetics, 116*, 272-278.

Nazrul Islam, M., Port, G. R., & McLachlan, A. J. (2001). The biology of *Lymnaea peregra* (Muller) (Gastropoda:Pulmonata:Basommatophora) with special emphasis to the effects of herbicides on its reproduction. *Online Journal of Biological Sciences, 1*, 532-540.

Palmer, C. G. S., Turunen, J. A., Sinsheimer, J. S., Minassian, S., Paunio, T., Lönnkvist, J., Peltonen, L., & Woodward, J. A. (2002). *RHD* maternal-fetal genotype incompatibility increases schizophrenia susceptibility. *American Journal of Human Genetics, 71*, 1312-1319.

Pyatt, F. B., Metcalfe, M. R., & Pyatt, A. J. (2003). Copper bioaccumulation by the freshwater snail *Lymnaea peregra*: A toxicological marker of environmental and human health? *Environmental Toxicology and Chemistry, 22*, 561-564.

Schellenberg, G. D., Dawson, G., Sung, Y. J., Estes, A., Munson, J., Rosenthal, E., Rothstein, J., Flodman, P., Smith, M., Coon, H., Leong, L., Yu, C.-E., Stodgell, C., Rodier, P. M., Spence, M. A., Minshew, N., McMahon, W. M., & Wijsman, E. M. (2006). Evidence for multiple loci from a genome scan of autism kindreds. *Molecular Psychiatry, 11*, 1049-1060.

Schilthuizen, M., & Davison, A. (2005). The convoluted evolution of snail chirality. *Naturwissenschaften, 92*, 504-515.

Suarez, B. K., Duan, J., Sanders, A. R., Hinrichs, A. L., Jin, C. H., Hou, C., Buccola, N. G., Hale, N., Weilbaecher, A. N., Nertney, D. A., Olincy, A., Green, S., Schaffer, A. W., Smith, C. J., Hannah, D. E., Rice, J. P., Cox, N. J., Martinez, M., Mowry, B. J., Amin, F., Silverman, J. M., Black, D. W., Byerley, W. F., Crowe, R. R., Freedman, R., Cloninger, C. R., Levinson, D. F., & Gejman, P. V. (2006). Genomewide linkage scan of 409 European-ancestry and African American families with schizophrenia: Suggestive evidence of linkage at 8p23.3-p21.2 and 11p13.1-q14.1 in the combined sample. *American Journal of Human Genetics, 78*, 315-333.

Weber, B., Hoppe, C., Faber, J., Axmacher, N., Fließbach, K., Mormann, F., Weis, D., Ruhlmann, J., Elger, C. E., & Fernández, G. (2006). Association between scalp hair-whorl direction and hemispheric language dominance. *Neuroimage, 30*, 539-543.

Wolman, D. (2005). *A left-hand turn around the world: Chasing the mystery and meaning of all things southpaw*. Cambridge, MA: Da Capo Press.

Wunderlich, R. C., & Heerema, N. A. (1975). Hair crown patterns of human newborns: Studies on parietal hair whorl locations and their directions. *Clinical Pediatrics, 11*, 1045-1049.

Ylisaukko-oja, T., Rehnström, K., Auranen, M., Vanhala, R., Alen, R., Kempas, E., Ellonen, P., Turunen, J. A., Makkonen, I., Riikonen, R., Nieminen-von Wendt, T., von Wendt, L., Peltonen, L., & Järvelä, I. (2005). Analysis of four neuroligin genes as candidates for autism. *European Journal of Human Genetics, 13*, 1285-1292.

In: Causes and Risks for Autism
Editors: A. C. Giordano et al.
ISBN: 978-1-60456-861-5

Chapter IV

"No Time Like the Present": Time Perception in Autism

***Melissa J. Allman**[*1] **and Iser G. DeLeon**[2]*
[1] The Kennedy Krieger Institute and Johns Hopkins University School of Medicine
[2] The Kennedy Krieger Institute and Johns Hopkins University School of Medicine

Abstract

The terms 'autism' and 'autistic' derive from the Greek word *autos* meaning self. This is appropriate to describing the autistic behavioral phenotype in which there is a pathological impairment in socialization and verbal and nonverbal communication, in addition to behavior and interests that are often highly restricted and repetitive (the triad; American Psychiatric Association, 1994). The autistic individual often appears isolated, and unable to make sense of the world around them. They often reveal an inability to predict and understand the behavior of others, and perceptions of the world remain fragmented and are not embedded into a coherent pattern or structure. Time is part of the fundamental intellectual structure in which we make sense of the events in our lives. 'Timing and time perception allow us to unite action sequences and events occurring separately in time, to adapt to

[*] Address correspondence to: Melissa J. Allman; Department of Behavioral Psychology ; Kennedy Krieger Institute; 707 N. Broadway; Baltimore, MD 21205

reoccurring situations, and to predicate behavior on what is expected to occur'. Timing and time perception are essential for adaptation and learning, memory and attention, cognitive development, and social synchrony and communication (see Meck, 2003). Firsthand accounts of people with autism often report a need to adhere to rituals or routines to compensate for a failure to predict events, and to their disorientation in time. They reveal a general lack of understanding about the passage of time, and appear stuck in the present. It is for these reasons that the issue of timing and time perception in autism is particularly intriguing. We will review empirical evidence that collectively suggests time perception may be disordered in autism, and postulate that fundamentally, a disturbed 'time sense' may contribute to features of the autistic behavioral phenotype.

Introduction

Time is an omnipresent feature of the human experience, and accordingly has been the subject of enquiry since the dawn of civilization (e.g., Saint Augustine, 397/398 AD) and the advent of modern psychology (e.g., James, 1890; see Fraisse, 1963 for a review of debate about time perception over 25 centuries). Time is so fundamental to our understanding of the world that it is difficult to imagine a world without it. Imagine if you could not represent the length of a minute, a year, or were unable to anticipate the impending change of a traffic light, or the length of a movie (how would you know that once started it would come to an end?). 'Our capacity to perceive temporal structure and our sensitivity to time on multifarious scales; from short durations (i.e., milliseconds and seconds) to long durations (e.g., hours and days), at perceptual, conceptual and linguistic levels, is central to our adaptive functioning, and all behavior is ultimately under the control of time' (Michon & Jackson, 1985). These authors assert that time "occupies a central position in our cognitive representations of reality" (1985, p. 3), and time is widely regarded as the most important dimension by which we make sense of the world (Navon, 1978).

Relatedly, there are anecdotal and clinical reports that individuals with autism experience difficulty in comprehending the passage of time, and this has a significant impact upon how they perceive the world (see Boucher, 2001). "Concepts of time have always puzzled and fascinated philosophers but most people are born with the ability to understand it in everyday terms. People with autistic disorders seem to lack this understanding to a degree that is markedly discrepant with their level of intelligence" (Wing, 1996: 89). Remarkably, very little empirical attention has been paid to examining time perception in autism (but see Boucher, Pons, Lind & Williams, 2007; Szelag, Kowalska, Galkowski & Poppel, 2004). This chapter will introduce elements in the typical development of time perception during infancy and childhood, and will review evidence that indicates these may be impaired in autistic individuals. It will speculate on how a disordered 'time sense' may contribute to the triad of autistic impairments.

Typical Development of Time Perception During Infancy

Developing an understanding of temporal structure has been postulated to depend upon any number of 'time experiences', a comprehensive discussion of which is beyond the scope of this chapter (but see Friedman, 1982; Lewkowicz, 1989; 1992; Poppel, 1978). Suffice to say here, that duration, synchrony, order, and 'past and present' appear to be key elements that lay the foundation for the perception and representation of time. Each of these elements will now be discussed in relation to autism.

Duration

The developmental psychology of time perception postulates that awareness and knowledge about temporal duration emerges from intrinsic biobehavioral rhythms and early action (Lewkowicz, 1989). The temporal regulation of rhythmic stereotypies displayed by infants (see Thelen, 1981) has been postulated to facilitate their adaptation to the temporal contingencies of their physical and social environments (Droit & Pouthas, 1992). Furthermore, especially in the early years, repetitive motor actions may themselves serve to measure the duration of events. Pouthas (1985) observed that when typically developing 10-24 month olds were required to withhold responding for a target delay (i.e., 5 s), they did so by adopting repetitive actions during the interval (i.e., they engaged in body-rocking or moved around the room in a certain fashion), but in older children between the ages of 4-7 yrs, manifestations of this 'behavioral clock' were significantly reduced (Pouthas & Jacquet, 1987). It appears that reliance upon motor actions to estimate duration may be supplanted (during development) by more cognitive processes. In a similar vein, mentally retarded children aged 6-17 (with an IQ between 29 & 48) can reproduce a given duration as accurately as typically developing children aged 3-6 years, if they are allowed to engage in repetitive behaviors (and have rhythmic structure) during the interval (see Fraisse, 1982). Perhaps then, a failure to acquire an understanding of duration and temporal contingencies accounts for the persistence of stereotypic behaviors in autistic individuals into adult life. This tenet is strengthened by empirical evidence (Szelag et al., 2004) that high-functioning children with autism (9-16 yrs) are severely impaired in their ability to reproduce target durations (between 1-5 s) compared to age-matched typically developing controls. These authors explicitly attribute such failures to deficits in the autistic internal timing system. An impaired understanding of duration is also revealed by the autistic "lack of awareness that an event, once started, will come to an end" (Wing, 1988, p. 88).

Synchrony

Synchrony is considered one of the first features of temporal experience to be differentiated by an infant (Lewkowicz, 1992). Here, we will refer to two aspects of synchrony: interactional synchrony (the ability to adapt to a temporal structure or coordinate with external events), and intermodal synchrony (the ability to perceive temporal synchrony in events occurring at the same time), both of which can be acquired during parent/infant interactions. It has been postulated that the temporal patterning of a parent's communication toward the infant is the most important aspect of these interactions for the child, and allows them to acquire an understanding of temporal expectancies and structure (Stern, Beebe, Jaffe & Bennet, 1977). An infant's ability to temporally coordinate his or her behavior with that of another person is key to preverbal and verbal interactions, in which turn-taking must be accurately timed (Trevarthen & Aitken, 2001). From birth, infants are typically able to synchronize with certain aspects of their parents' communication (Malloch, 1999), and exquisite reciprocal behavioral rhythms and regularities are observed during parent/infant interactions (for details, see Lester, Hoffman & Brazelton, 1985). However, asynchronous social coordination (during these interactions) has been retrospectively observed in 11-month-old autistic infants (Trevarthen & Daniel, 2005; see also Kubicek, 1980). Moreover, the discrimination of temporal synchrony between intermodal events (i.e., the sight and sound of the parent's speech) during parent/infant interactions "may be the first step in developing a capacity to discriminate more complex and specific forms of language" (Bebko, Weiss, Demark & Gomez, 2006, p. 96). These authors report that autistic children (aged 4-6 yrs) reveal atypical responding to mutimodal temporal asynchrony with language-specific stimuli. Collectively, these findings lend support to the suggestion that an understanding of temporal synchrony may be impaired in autistic individuals.

Order and 'Past and Present'

The order of successive or serial events is a fundamental aspect of temporal structure, and also provides information as to the causal relationships between events. Harner (1982) asserts that two types of seriation exist. The first is the relative position of two events on a time continuum (i.e., one event precedes the other). This form of knowledge about temporal order is likely spared in autism, as paired-associate learning and linear sequencing of successive items is typically intact (Kanner, 1943; Minshew, Goldstein & Siegal, 1997) and is commonly evidenced by echolia, and rote learning for phrases or songs (Boucher, 2001). This type of 'temporal stringing' has been argued to depend upon linear (or circular) visual representations of temporal order (Friedman, 1990). This is interesting in relation to autism, as Temple Grandin, one high profile autistic individual reports "my mind is like a...quick access videotape. But [in order to remember an aspect of an event] I have to play the whole part—no fast forward" (Sacks, 1995, p. 269). However "it seems that people with autistic disorders have severe problems coping with sequential events that have no independent, concrete existence" (Wing, 1988, p. 89), so a child may have difficulty understanding the concepts of 'yesterday', and 'tomorrow', unless these are concretized by showing them a calendar and the relevant dates. This is evidenced by the popularity and success of picture schedules (of temporally sequenced events) in the training and treatment of children with autism (e.g., Lalli, Casey, Goh & Merlino, 1994; MacDuff, Krantz & McClannahan, 1993). The second aspect of seriation is the position of events in the overall time continuum; although two events will maintain a consistent relation to one another, their inclusion in the before or after; or past, present and future categories of experience is transitory (e.g., tomorrow becomes yesterday). A sensitivity to this more complex form of sequential structure is considered crucial for action knowledge, object use, drawing inferences from others actions, and planning one's own behavior (Baldwin, Baird, Saylor & Clark, 2001; Zalla, Labruyere & Georgieff, 2006). This type of hierarchical temporal coding is likely deficient in autism (see Boucher, 2001, p. 113). For instance, it has recently been reported (Boucher et al, 2007) that children with autism (7-16 yrs) reveal marked impairments in diachronic thought. That is, they are unable to i) think about past or future stages of current situation,

ii) understand that things can change or evolve over time but are still the same thing, and iii) that successive events are part of a unitary process (see also Montangero, 1992).

How a Disorder of Time Perception May Contribute to the Autistic Behavioral Phenotype

It is difficult to know with any degree of certainty how a disordered sense of time might impact behavioral and cognitive function, but is easy to speculate. That the products of this speculation so closely resemble features of the autistic behavioral phenotype is particularly striking.

Restricted and Repetitive Behaviors

Janet (1928) noted that one of our earliest experiences with time (duration) arises during periods of waiting, when there is an imposed delay between our desires and their satisfaction. An inability to wait represents a deficit in linking the passage of time with ongoing activities, and is a common problem for autistic children and adults (Wing, 1988). "Impatience is common in all young children but in people with autistic disorders it can continue for years, even into adult life" (1988, p. 88). Recall that in one study (Pouthas, 1985), waiting in young children was facilitated by their adoption of repetitive behaviors (e.g., stereotypies) that functioned to parse the delay interval. That stereotypies can function as a 'behavioral clock', suggests that a failure to understand the passage of time (duration) may account for the persistence of certain repetitive behaviors in autism. Stereotypies are typically produced in repeating cycles, and may be seperated by (often short) intervals in time—continually measuring intervals in a repeating cycle requires less attentional resources (Lewis & Miall, 2003), and so repetitive motor behaviors may be a particularly effective time-parsing stratergy for autistic individuals, and might function to concretize and reduce the stressor of an imposed disorientation in time. Peeters and Gillberg (1999, p. 87) report that "most people with autism feel lost in a sea of time…they will often try to develop routines and rituals by way of compensation. They want all activities to be undertaken in the same sequence every day…and if the sequence of activities changes on a certain day, then they have behavior problems" [which can include rhythmic lower-order motor behaviors, e.g., head-banging, self-injury]. To reiterate the main point here, an autistic impairment in the perception of duration may be compensated for by the production of repetitive motor behaviors (such collateral behaviors are often observed during superior temporal performance in animal studies), and an overreliance upon intact abilities, such as sequencing and order, and the stringing together of temporal units of perseveration or habits.

One very interesting possibility was considered by Boucher (2001). In her own words, "try to imagine periods of time longer than the lifetime of the universe…in fact, one cannot imagine a period of time longer than the lifetime of the universe except by thinking of a temporal succession of universes with cumulative lifetimes" (2001, p. 121). She suggests that there may be a close correspondence between the length (and complexity) of repeating behavioral units (e.g., stereotypies, rituals) and the ability to imagine extended time frames in autism. As shorter and less complex stereotypies are usually observed with lower-functioning autistic individuals, and more complex, rigid routines are observed in those who are higher functioning, it follows that the ability to perceive duration might account for quantitative and qualitative differences in repetitive behaviors across the autistic spectrum (Boucher, 2001).

Language and Social Communication

Selective impairments in the autistic child's' ability to temporally coordinate to their social environment and to represent the temporal structure of their social and physical world, may explain the lack of interest in social interaction from the first year of life which parents of autistic children typically report (Wing, 1988). It may also contribute to these individuals retreating into their own 'inner world' (aloofness), and explain their preoccupation for repeating or rhythmical sensory stimulation (e.g., Grandin, 2005). Autistic deficits in acquiring knowledge from these early 'time experiences' (e.g., synchrony, duration) might produce a cascade of other autistic deficits (a full discussion of which is beyond the scope of this chapter). For example, we have seen that "in speech perception, temporal factors such as synchrony, duration, rate and rhythmic structure play an important role in the integration of the visible and audible aspects of the signal" (Lewkowicz, 1992, p. 34). Generative language production and comprehension is heavily dependent upon a multitude of temporal competencies, and with its immanent references to time (e.g., past, present and future tenses), language may become an irreconcilable code for an individual who lacks temporal fluency, as we postulate in autism.

The Autistic Perceptual Experience

Our subjective experience of duration is fallible, and can be influenced by a variety of factors (such as the content of the interval, drugs, and body temperature, e.g., see Meck, 2003). The popular phrase 'time flies when you're having fun' is testament to this quality of duration perception. There is also a debated phenomena known as 'time dilation'. This refers to the sensation that time can appear to slow down, or pass by in slow motion (particularly during periods of heightened arousal). When Alice fell down the rabbit hole at the beginning of her adventures in Wonderland, "either the well was very deep, or she fell very slowly, for she had plenty of time as she went down to look about her, and to wonder what was going to happen next" (L. Carroll, 1992). We might be familiar with a similar experience when our car starts to skid off the highway at speed. This phenomenon of time 'warping' is believed to be the product of an increase in the speed of the internal timing apparatus (and is currently being investigated by D. M. Eagelman and colleagues). It may be biologically adaptive, as it appears to produce a hypersensitivity to sensory events, and prompts an elemental rather than configural processing bias that serves to facilitate decision-making ability. During these moments, the individual is also 'stuck' in the subjective time present. In light of the fact that autistic individuals appear to experience qualitative differences in sensory perception (Grandin, 2005) it is possible that the subjective experience of time is more mercurial to intrinsic and/or extrinsic variables in people with autism; with the 'speed' or function of the internal timing apparatus being different or more variable in these individuals. Within this conceptual framework, lower-order rhythmic motor behaviors may function to regulate and stabalize the subjective perception of duration. Furthermore, if the perception of duration is anomalous between different sensory modalities in autism (i.e., visual and auditory; see Penney, Gibbon & Meck, 2000) then this may produce problems with intersensory integration (and intermodal synchrony), and the binding of external inputs into meaningful information (e.g., see Brock, Brown, Boucher & Rippon, 2002).

Conclusion

The interest in understanding time perception in normal and patient populations (including conditions co-morbid with autism) is growing rapidly (e.g., Toplack, Dockstader & Tannock, 2006), yet the paucity of empirical data relevant to autism is particularly striking (Boucher et al., 2007) given anecdotal reports that 'whatever *it* is that typically developing individuals posses that gives them a sense of timing we, as individuals with autism certainly lack it' (Lawson, 2001, p. 43). This chapter, and those studies which it cites, have elucidated how deficits in the experience of time may contribute to the autistic behavioral phenotype (the triad; APA, 1994). Pursuing this line of enquiry may enhance our understanding of this perplexing disorder, and advance extant interventions and strategies designed to concretize the passage of time for autistic individuals. An autistic preoccupation with timetables, clocks and calendars is common and may be particularly useful in helping autistic individuals to understand time (Wing, 1988). For example, it would be of particular interest to examine whether devices such as the Time Timer™ are effective in reducing repetitive behaviors during periods of waiting. Given that sensitivity to the temporal parameters of experience can be evidenced within the first year of life, it may also be worthwhile to incorporate assessments of temporal competence into early-year studies of at-risk autistic infants, in the hope that they may provide some predictive power as to later diagnosis.

References

American Psychiatric Association (1994). Diagnostic and Statistical Manual of Mental Disorders (4th Ed.). Washington, DC: American Psychiatric Press.

Augustinus: Confessions. Book 11. 397/398.

Baldwin, D. A., Baird, J. A., Saylor, M. M. & Clark, M. A. (2001). Infants parse dynamic action. *Child Development.* 72, 708-717.

Bebko, J. M., Weiss, J. A., Demark, J. L., & Gomez, P. (2006). Discrimination of temporal synchrony in intermodal events by children with autism and children with developmental disabilities without autism. *Journal of Child Psychology and Psychiatry.* 47, 88-98.

Boucher, J. (2001). 'Lost in a sea of time': Time-parsing and autism. In Hoerl, C. & McCormack, T. (Ed.). *Time and memory.* (pp. 111-135). Oxford: Oxford University Press.

Boucher, J., Pons, F., Lind, S. & Williams, D. (2007). Temporal cognition in children with Autistic Spectrum. *Journal of Autism and Developmental Disorders.* epub ahead of print.

Brock, J., Brown, C. C., Boucher, J & Rippon, J. (2002). The temporal binding deficit hypothesis of autism. *Development and Psychopathology*. 14, 209-224.

Carroll, L. (1992). Alice's adventures in Wonderland & through the looking glass. Hertfordshire: Wandsworth Editions Limited.

Charman, T., Baron-Cohen, S., Swettenham, J., Baird, G., Drew, A., & Cox, A. (2003). Predicting language outcome in infants with autism and pervasive developmental disorder. *International Journal of Language and Communication Disorders*. 38, 265-285.

Droit, S., & Pouthas, V. (1992). Changes in temporal regulation of behavior in young children: From action to representation. In Macar, F., Pouthas, V., & Friedman, W. J. (Eds.). *Time, action and cognition: Towards bridging the gap*. (pp. 45-53). The Netherlands: Kluwer Academic Publishers.

Fraisse, P. (1963). Introduction. In Fraisse, P. (Ed.). *The psychology of time*. (pp. 1-15). New York: Harper & Row.

Fraisse, P. (1982). The adaptation of the child to time. In W. J. Friedman (Ed.). *The developmental psychology of time*. (pp.113-140). New York: Academic Press.

Friedman, W. J. (1982). Introduction. In W. J. Friedman (Ed.). *The developmental psychology of time*. (pp. 1-11). New York: Academic Press.

Grandin, T. & Johnson, C. (2005). Animals In Translation: Using The Mysteries Of Autism To Decode Animal Behavior. New York: Harvest Books.

Harner, L. (1982). Talking about the past and the future. In W. J. Friedman (Ed.). *The developmental psychology of time*. (pp. 141-169). New York: Academic Press.

James, W. (1890). *The principles of psychology*. Vol 1 (pp. 605-642). New York: Holt.

Janet, P. (1928). L'evolution de la memoire et de la notion de temps. Paris: Chahine.

Kanner, L. (1943). Autistic disturbances of affective contact. *Nervous Child*. 2, 217-250.

Lawson, W. (2001). Understanding and working with the spectrum of autism: An insider's view. Jessica Kingsley Press, London.

Lester, B. M., Hoffman, J., & Brazelton, T. B. (1985). The rhythmic structure of mother-infant interaction in term and preterm infants. *Child development*. 56, 15-27.

Lewkowicz, D. J. (1989). The role of temporal factors in infant behavior and development. In I. Levin, & D. Zakay (Eds.). Time and human cognition. *A life-span perspective. Advances in Psychology.* 59. (pp. 9-62). The Netherlands: Elsevier Science Publishers.

Lewkowicz, D. J. (1992). The development of temporally-based intersensory perception in human infants. In F. Macar, V. Pouthas, & W. J. Friedman (Eds.). Time, action and cognition. *Towards bridging the gap.* (pp. 33-43). The Netherlands: Kluwer Academic Publishers.

Lewis, P. A. & Miall, R. C. (2003). Distinct systems for automatic and cognitively controlled time measurement: Evidence from neuroimaging. *Current opinion in neurobiology.* 13, 250-255.

Lalli, J. S., Casey, S., Goh, H. & Merlino, J. (1994). Treatment of escape-maintained aberrant behavior with escape extinction and predictable routines. *Journal of Applied Behavior Analysis.* 27, 705-714.

Kubicek, L. F. (1980). Organisation in two mother-infant interactions invloving a normal infant and his fraternal twin brother who was later diagnosed as autistic. In T. Field (Ed). *High Risk Infants and Children.* (pp. 99-110). New York: Academic Press.

MacDuff, G. S., Krantz, P. J. & McClannahan, L. E. (1993). Teaching children with autism to use photographic activity schedules: Maintenance and generalization of complex response chains. *Journal of Applied Behavior Analysis.* 26, 89-97.

Malloch, S. N. (1999). Mothers and infants communicative musicality. *Musicæ Scientiæ.* 29-57.

Marquis, D. P. (1941). Learning in the neonate. *Journal of Experimental Psychology.* 29, 263-282.

Meck, W. H. (2003). Functional and neural mechanisms of interval timing (pp. xvii-xli). Florida: CRC Press LLC.

Michon, J. A & Jackson, J. L. (1985). Introduction: The psychology of time. In J. A. Michon & J. L. Jackson (Eds.). *Time, mind and behavior.* (pp. 2-17). New York: Springer-Verlag.

Minshew, N. J., Goldstein, G., & Siegal, D. J. (1997). Neuropsychologic functioning in autism: Profile of a complex information processing disorder. *Journal of International Neuropsychological Society.* 3, 303-316.

Montangero, J. (1992). The development of a diachronic perspective in children. In F. Macar, V. Pouthas, & W. J. Friedman (Eds.). Time, action and cognition. *Towards bridging the gap.* (pp. 55-65). The Netherlands: Kluwer Academic Publishers.

Navon, D. (1978). On a conceptual hierarchy of time, space and other dimensions. *Cognition.* 6, 223-228.

Peeters, T., & Gillberg, C. (1999). Autism: Medical and educational aspects. London: Whurr Publishers.

Penney, T. B., Gibbon, J. & Meck, W. H. (2000). Differential effects of auditory and visual signals on clock speed and temporal memory. *Journal of Experimental Psychology: Human Perception and Performance*. 26, 1770-1787.

Poppel, E. (1978). Time perception. *Handbook of sensory physiology.* VIII (pp. 713-729). Germany: Springer-Verlag.

Pouthas, V. (1985). Timing behavior in young children: A developmental approach to conditioned spaced responding. In J. A. Michon & J. L. Jackson (Eds.). *Time, mind and behavior.* (pp.100-109). New York: Springer-Verlag.

Pouthas, V. & Jacquet, A. Y. (1987). A developmental study of timing behavior in 4 ½ and 7-year-old children. *Journal of Experimental Child Psychology.* 43, 282-299.

Sacks, O. (1995). An anthropologist on Mars: *Seven paradoxical tales.* London: Picador.

Stern, D. N., Beebe, B., Jaffe, J. & Bennet, S. L. (1977). The infant's stimulus world during social interaction: A study of caregiver behaviors with particular reference to repetition and timing. In H. R. Shaffer (Ed.). *Studies of mother-infant interaction.* (pp. 177-204). New York: Academic Press.

Szelag, E., Kowalska, J., Galkowski, T., & Poppel, E. (2004). Temporal processing deficits in high-functioning children with autism. *British Journal of Psychology.* 95, 269-282.

Thelen, E. (1981). Rhythmical behavior in infancy: An ethological perspective. *Developmental Psychology.* 17, 237-257.

Toplak, M. E., Dockstader, C. & Tannock, R. (2006). Temporal information processing in ADHD: Findings to-date and new methods. *Journal of Neuroscience Methods.* 15-29.

Trevarthen, C., & Aitken, K. J. (2001). Infant intersubjectivity: Research, theory and clinical applications. *Journal of Child Psychology and Psychiatry.* 42, 3-48.

Trevarthen, C., & Daniel, S. (2005). Disorganized rhythm and synchrony: Early signs of autism and Rett syndrome. *Brain & Development.* 27, S25-S34.

Wing, L. (1996). The Autistic Spectrum. London: Constable.

Zalla, T., Labruyere, N. & Georgieff (2006). Goal-directed action representation in autism. *Journal of Autism and Developmental Disorders.* 36, 527-540.

In: Causes and Risks for Autism
Editors: A. C. Giordano et al.
ISBN: 978-1-60456-861-5

Chapter V

Autism and Dysphasia: A Physio-Pathological Approach

Claudia Talero-Gutierrez
Neuroscience Unit; Neuroscience Group (NEUROS)
School of Medicine; Universidad del Rosario; Bogotá, Colombia

Abstract

Language abnormalities are critical in diagnosing autism, including the absence of or severe language delay, inability to initiate or sustain a conversation and the use of stereotypic or repetitive language. Symptoms related to language may vary in intensity; however, the common denominator (i.e., an inability to establish effective communication) is ever present in these patients.

Other diagnostic criteria address deficiency in or lack of social interaction, reflected in abnormal play behavior or in the manipulation of elements having inappropriate, repetitive patterns of a non-functional ritualistic nature.

Children suffering from autism generally present no major alterations of early motor development during the first and second years of life and the child is able to sit, crawl and walk within the expected time-frames. Retrospective studies of films of autistic children have shown subtle differences in object manipulation, visual attention to social stimuli as well as in smiling and exploration. However, the alarm signal (and the one usually triggering consultation) is the absence of or defective language development.

The above observations indicate that the central alteration in an autistic child resides in the ability to relate to others, particularly involving verbal and non-verbal communication, difficulties in establishing social contact and an inability to detect the intentions of others.

Neuro-anatomical studies and functional magnetic resonance imaging have shown structural variations and alterations involving the cerebellum, as well as temporal and frontal lobes. One of the striking features in these patients from the clinical viewpoint is the absence of alerting reaction and response to the human voice, while the orientating reflex to environmental sound stimuli remains normal. Dysphasia may range from lack of recognizing basic units or phonemes to problems in integrating a particular word with its meaning. The clinical picture becomes more complex as a child grows older and more participation in activities requiring social interaction is expected.

Early recognition and appropriate treatment of difficulties with processing social information provides a basis for treatment methodology that is often able to change these patients' worsening clinical course.

Introduction

Severe delay in the appearance of language or the absence of its development, an inability to initiate or maintain a conversation and the use of stereotyped or repetitive language are some of the diagnostic criteria for autism. Added to language disorders, one finds alterations in interpersonal relationships and an inadequate response to stimuli, poor eye-contact and repetitive motor behavior, which may be totally or partially present.

Symptoms related to language may vary in intensity; however, they have a common denominator which is an inability to establish effective communication.

Neurobiological studies on autism during the last few years have contributed towards elucidating the disease's etiological factors and neuro-anatomical correlates leading to a physio-pathological explanation of the different difficulties observed in these patients.

Given that communication deficiencies are a fundamental part of such clinical picture, the present commentary is centered on an approach to the physiopathology of language disorder in these children and its relationship to other symptoms of autism. Starting from this viewpoint, it may be understood that an early intervention orientated towards developing communication strategies may modify the course of a child's development and improve his/her prospects of adapting to society.

Language in Diagnostic Criteria Regarding Autism

Diagnostic criteria for autism, for both DSM IV [1] and CIE 10 [2], are based on three groups of symptoms related to social interaction, communication and behavior. Limitations may be observed in verbal and non-verbal communication ability as well as in social relationships. In such patients language may be absent or severely delayed; those who manage to develop some language do it inadequately and present difficulties in understanding [3].

The natural history of the evolution of children suffering from autism is characterized by the symptoms most noticeably observed in their communication (or lack of it). The problem of language is often the first symptom presented in autism [4]. Little gurgling and limited vocalization may be found during the lactation period compared to other babies and, on some occasions, there is no language development (if there is, it is neither understandable nor expressive). When language develops, it is usually altered in form, use and structuring. It might be repetitive, metaphorical, rhetorical or echolalic and the speaker may refer to him/herself in the third person. The voice is generally monotonous and the tone might be shrill. Prosody might be altered, producing a characteristic "robotic talk." However, such children are able to repeat, quite literally, a discourse heard with very close imitation, in some cases with identical characteristics of the voice, rhythm and intonation of the one being imitated. Some do not acquire language and seem to understand very little of what is said to them.

These children have play limitations and focus themselves on manipulating objects; they are not creative regarding representation via symbolic play. They linger on aspects of objects such as smell, taste, color or brilliance, texture or movement. They seem to perceive parts of objects and people in isolation. Given the previous characteristics, interrelationships are difficult with other children [5].

Autism is also defined by altered social interaction and stereotyped repetitive behavior and interests. Altered communication ability may generally vary from non-functional language to alterations in phonological processing, semantics, syntax and vocabulary compared to their peers of the same age [6].

Alterations in the development, use, prosody and other aspects of language can also be found in other clinical pictures included in the category of generalized development disorders. Such clinical pictures can have different degrees of complexity ranging from apparently adequate language use (as in Asperger's syndrome) to severe deterioration in communicative abilities (i.e. Rett's syndrome).

Neuro-Anatomical Studies

The structural and functional characteristics of the brains of those suffering from autism have been the subject of careful study and differences have been identified in some structures and variations in activating specific areas of the central nervous system (CNS), especially when people have to face different types of tasks.

Since Kanner described the first cases, observations have been made regarding the relative increase in cranial perimeter which, according to Courchesne *et al.*, in work with different age groups, has been confirmed as an increase in relative size during infancy, followed by a period of abnormal slowing down of growth so that no significant differences in cerebral volume are found in older children and adolescents [7].

Neuro-anatomical studies have been published since 1985 showing structural differences in the cerebral cortex of autistic people and contributions were made during the 1990s to clarifying different cerebral structures' morphological characteristics in these patients.

The first studies found reduced number of Purkinje cells in the vermix and cerebellar hemispheres. Smaller size but increasing numbers of neurons have been described in the inferior olive [8]. Small neurons, poor lamination of the anterior cingulate cortex and cortical dysgenesis have been found in 4 out of 6 autistic people diagnosed in life [9] [10]. Pathological study has revealed reduced cortexes, high neural density, the presence of neurons in the molecular layer, an irregular laminar patterns and a lack of defined limits between grey and white matter. The foregoing microscope pathology findings have been found in the cortex, cerebellum and cerebral stem, suggesting errors in neural growth and cell migration [10].

The configuration of the so-called minicolumns described by Mountcastle [11] as being anatomical and functional units in the cortex's smallest level of vertical organization have been investigated in autism, smaller and less compact minicolumns having been found in nine autistic subjects compared to nine controls and in two patients having Asperger's syndrome compared to 18 subjects as control [12].

Classical neuro-pathological observation regarding autism have revealed increased cell group density, lesser neuron size in the limbic system, reduced number of Purkinje cells in the cerebellum and evidence of cortical dysgenesis or altered migration [8].

The corpus callosum, an index for neural connections between cerebral regions, has also been studied. Overall reduction in size of the corpus callosum, involving in particular the anterior regions has been described, indicating lesser inter-hemisphere neuronal connections. Greater reduction has been found in the genu of the corpus callosum corresponding to prefrontal cortex projections [13]. These findings are consistent with cognitive, behavioral and neurophysiological evidence of dysfunction in the frontal lobe related to deficit in executive function, spatial work memory and the ability to suppress inappropriate responses [14].

Brain imaging studies have revealed reduced grey matter density in Brodmann's area 45 in the left inferior frontal gyrus in adults suffering autism, as well as asymmetry in the volume of the inferior frontal cortex and the pars opercularis. Left temporal planum volume has been found to be reduced in adults suffering from autism compared to that of normal adults whilst greater left side volume has been found in autistic children compared to controls and abnormalities in superior temporal gyrus in most autistic children. Functional brain imaging studies have generally revealed abnormalities in autistic individuals' temporal activation [6] [7] [13] [15]. The cerebellum has been implicated in multiple functions such as fixing attention, language, procedural memory and non-motor learning. Abnormal activation has been reported in autistic people regarding attention and motor tasks [16]. It has been proposed that cerebellum alteration in autism could limit children's ability to learn predictive relationships between sequences of events and altered attention.

Language Development and Dysphasia

For language to develop adequately, the CNS must be able to process different and multiple sensory afferent signals and integrate them in order to properly organize the perceived information. Moreover, speech requires the capacity to structure and program the expression of a determined statement which will transmit the message which is being expected to communicate. Likewise, some sensory receptors, especially unscathed auditive and visual ones, are needed which can capture and transmit the received information to the CNS. Verbal expression requires regulator centers and nervous system programmers, structurally suitable phono-articulator organs allowing modulation of the sounds of the voice for emitting words and a motor innervation mechanism leading to programming the fine movements which must be made when emitting a word.

All the forgoing must be immersed in a propitious, enriching setting, allowing a subject to develop experiences that lead to the construction of communication codes valid in a person's surrounding environment.

If the previous model is born in mind, then language development disorders can be classified as having a central, sensorial, motor and coordination, peripheral and/or environmental origin.

Although developmental language disorders compromise auditory capacity, they do not block the communicative intention of a child who must be understood by signals and gestures. If such disorders arise from motor alterations and affect phono-articulator organs, then difficulties will be observed in expressive language aspects, especially in articulating the different phonemes in such a way that talking is difficult to understand in some cases or some phonemes are badly pronounced. When failures in environmental stimulation occur, then language development is delayed; this could be reversed be becoming immersed in an environment enriched by linguistic information or by specific language stimulation therapy.

Central language disorders (i.e., those that arise from central nervous system involvement) may have different levels of severity.

There can be different types of language development delay. Some are called simple delays, being presented during different stages of development but resulting in slight to moderate delay when compared with the events observed in peers; there are also delays secondary to cognitive dysfunction, or complex ones in which control of language processing is affected and children perform below the level for their age-group in tasks requiring phonological, syntactical or semantic processing, in spite of being in a suitable environment and possessing normal intelligence. This type of language development disorder has been associated with the FOXP2 gene [17] [18].

Likewise, some delays of language development involve children with adequate hearing and motor development, together with rich environmental stimulation, but who are not alert to language and/or to gesture, waving and signals; such afferences are not processed, and these behaviors are more in line with aphasia than with auditive agnosia, even though some authors have proposed a primary auditive agnosia in autistic children [5] [19]. Such aphasia (also called dysphasia due to the characteristic alterations during language development) could have alterations limited to areas related to the processing of verbal afferences to the temporal lobe, in which case it will mainly affect understanding and limit the compromise to Wernicke's area to a greater extent. Dysphasia will be expressive in cases where, even though there is understanding and the ability to capture a message received from a speaker, there is an impossibility to respond to this message and it will be mixed when the two processes become compromised. The fact of responding to stereotyped language can reveal the possibility of establishing connection circuits between verbal afferences and efferences without information having been integrated in processing areas such as Wernicke's area. This is the case of children who repeat or talk automatically, without content or who engage in echolalia [20] and in those children who spontaneously and automatically develop the ability to read without understanding the content of a written text [21].

Communication disorders, according to DSM IV can be categorized as expressive language disorder, expressive receptive language mixed disorder, phonological disorder, stammering and non-specified communication disorders [1]. Of the foregoing, mixed understanding-expressive disorder corresponds to what is called mixed dysphasia in other classifications (dysphasia in the sense that it constitutes language difficulty or severe alteration in acquiring it). In this picture of development, such dysphasia is comparable to acquired aphasia in that the greatest compromise is found in the ability to decode an auditory message that is heard but whose verbal expression cannot be programmed.

Physiopathological Proposal

The model proposed here does not ignore the presence of other components in autism contributing towards the clinical picture; however, it specifically focuses on communication given that it is the predominant symptom of autism and the pillar for diagnosis.

During the first and second years of life, the autistic child's motor development is found to be within expected parameters and does not present alterations in holding up the head, sitting up, crawling or walking. Their fine motor abilities also show no compromise, even though they could present some motor behavior which is slightly different to that of their peers when manipulating objects, a fact which could obey to the characteristics of their perception more than to a specific alteration of their manipulative ability. The characteristics of the first year of life have been studied by filming autistic children and comparing them to peers who have developed adequately, finding relatively less visual attention and vocalization, as well as placing less emphasis on detail when exploring objects. These subtle changes hamper the diagnosis of the clinical picture during the first year of life. Most parents of autistic children detect some type of abnormality around 24 months of age and request a consultation, generally due to the anxiety generated by the child's reduced auditory ability and the lack of language development [21].

Careful clinical observation reveals normal responses to sound stimuli from the environment with lack of alertness to auditory signals from the human voice, lack of recognition of their own names and delay in developing understanding and language expression.

Autism encompasses a wide variability of responses and (depending on the degree of compromise) autistic children may present themselves as being more or less cut off socially. It is important that the children's other types of behavior should be considered when evaluating the characteristics of their attempts to establish communication. The absence of language in a deaf child does not impede seeking communication by other routes and establishing codes via gesture and signals leading to establishing contact with their parents and family.

When hypoacusis is diagnosed early, a child can receive treatment before developing other forms of communication. Likewise, children present with alterations in expressive language development (from simple delay in developing language up to verbal praxis alterations) they usually exhibit behaviours that indicate the appropriate connection with the environment and reveal their understanding ability, although their expression is thereby limited. This is not so for children presenting mixed alterations having understanding and expressive components in which their ability to understand and to express are compromised. These children and those called autistic have difficulty in decoding spoken messages and encoding verbal and non-verbal messages in common in their attempts to communicate [4] [22].

Cases are found within the same spectrum of so-called autism in which compromised communication is associated with more or less severe forms of isolation. According to such variability, there are cases of autistic children in which the lack of contact becomes severely extended, even affecting their ability to establish visual contact. Such children are probably those in which altered neurodevelopment more severely involves more extensive areas of the CNS. There are other cases in which blocking and lack of response to the human voice are notable; however, visual and tactile contact is achieved without triggering intolerance, which could correspond to their inadequate ability to process signals from the environment. In some autistic children, the clinical picture shown from the language point of view lies in their inability to understand and process the sounds of language, whilst sounds from the environment and even musical sounds alert them and in some cases trigger their fascination.

Neuro-anatomical studies and image studies have revealed structural variations of temporal and frontal regions involved in processing language [8] [9] [10] [12] [13] [14]. When the temporal cortex is involved there are alterations in processing afferent information thereby hampering a child's capacity to recognize and to integrate the sounds of the human voice with words and with the significance so transmitted, in such a way that there is total absence of response to triggering and understanding language. Altered neurodevelopment probably involves the neural circuits processing verbal information from the most basic stages of recognition of the human voice so that a child cannot interpret language. The same defect, this same difficulty in processing, may explain why, on occasions, a child can repeat words, phrases outside their context or engage in echolalia, without understanding the language which is being used. Children can sometimes emit words and phrases out of context as if a direct connexion had been established between the circuits receiving this information and the routes leading to their repetition but without having been processed by the integration circuits associating the sounds with their meaning (Wernicke's area).

Frontal region dysfunction will be implicated in altered verbal expression, from being deaf and mentally retarded to flaws in syntactical and grammatical organization, thereby making speech (when there is any) become stereotyped and de-structured or out of context. There will be no programming of pre-verbal and verbal communicative intention or it will be altered.

Given the language alterations presented in this model (being coherent with evolutionary dysphasia), it can be considered that there is an underlying alteration of this type in the picture of autism accompanied by severe language compromise.

Conclusion

Managing a child presenting a clinical picture compatible with the DSM IV classification of autism must be done by a team of rehabilitation professionals establishing an intervention program and family support.

If this springs from the physio-pathological proposal so described, then treatment should be orientated towards developing communication, as well as implementing other abilities compromised as a consequence of alterations in other information processing areas in the CNS.

Even though improvements have been described in some autistic children's stereotyped and disruptive behavior by using pharmacological treatment, there is currently no specific medical treatment for autism [23] [24].

Considering that a child diagnosed as being autistic is suffering from a primary communication disorder, then therapy orientated towards stimulating both cognitive and expressive language development can provide promising results when begun early enough and when there is suitable collaboration by all the people forming part of the environment surrounding an autistic child.

Language and occupational therapy must be carried out jointly to establish routines and to arrange treatment contents according to a particular child's evolutionary stage.

From the point of view of processing auditive information, amplifying sound by selecting specific frequencies (according to the observation of comfort expressed in the child's behavior) could be a method aiding the development of auditive perception and slowly integrating the sounds of language within a significant context.

All activities carried out during therapy must be done in play-based environments initially containing few stimuli and which, little-by-little, become enriched by greater elements. Planning therapy with due respect for these children's susceptibility to excess stimuli will lead to slowly developing the ability to recognize the keys to communication which a child can integrate.

Last but not least, classifying autism as being primarily dysphasic language usually changes parents, educators and rehabilitators' expectations, reduces the distress produced by a diagnosis of autism and frequently improves a patient's development perspectives.

References

[1] American Psychiatric Association. Diagnostic and Statistical Manual of Mental Disorders. Fourth Edition. Washington, DC: American Psychiatric Association; 1994.

[2] World Health Organization. International Classification of Diseases (ICD-10). Available from: URL: [http://www.who.int/classifications/icd/en/]
[3] Frith, U. Mind blindness and the brain in autism. *Neuron*, 2001; 32: 969-979.
[4] Kjelgaard, MM. Tager-Flusberg, H. An investigation of Language Impairment in Autism: Implications for Genetic Subgroups. *Lang. Cogn. Process*. 2001; 16 (2-3): 287-308
[5] Holguin, JA. El autismo de etiología desconocida. *Rev. Neurol.* 2003; 37(3): 259-266.
[6] De Fosse, L; Hodge, SM; Makris, N; Kennedy, DN; Caviness, VS; McGrath, L; Steele, S; Ziegler, DA; Herbert, MR; Frasier, JA; Tager-Flusberg, H; Harris, GJ. Language-Association Cortex Asymmetry in Autism and Specific Language Impairment. *Ann. Neurol.* 2004; 56(6):757-766
[7] Courchesne, E; Karns, CM; Davis, HR; Ziccardi, R; Carper, RA; Tigue, ZD; Chisum, HJ; Moses, P; Pierce, K; Lord, C; Lincoln, AJ; Pizzo, S; Schreibman, L; Haas, RH; Akshoomoff, NA; Courchesne, RY. Unusual brain growth patterns in early life in patients with autistic disorder: an MRI study. *Neurology*. 2001; 57(2): 245-54
[8] Palmen, SJ; van Engeland, H; Hof, PR; Schmitz, C. Neuropathological findings in autism. *Brain*. 2004; 127(Pt 12): 2572-2583
[9] Kemper, TL; Bauman M. Neuropathology of infantile autism. *J. Neuropathol. Exp. Neurol.* 1998; 57(7): 645-652
[10] Bailey, A; Luthert, P; Dean, A; Harding, B; Janota, I; Montgomery, M; Rutter, M; Lantos, P. A clinicopathological study of autism. *Brain*. 1998; 121 (Pt 5): 889-905.
[11] Mountcastle, VB. The columnar organization of the neocortex. *Brain*. 1997;120 (4): 701-722
[12] Casanova, MF; Buxhoeveden, DP; Brown, C. Clinical and macroscopic correlates of minicolumnar pathology in autism. *J. Child Neurol.* 2002;17(9):692-695
[13] Just MA, Cherkassky VL, Keller TA, Kana RK, Minshew NJ. Functional and anatomical cortical underconnectivity in autism: evidence from an FMRI study of an executive function task and corpus callosum morphometry. *Cereb. Cortex*. 2007; 17:951-61.
[14] Hardan, AY; Minshew, NJ; Keshavan, MS. Corpus callosum size in autism. *Neurology*. 2000; 55(7):1033-1036

[15] Penn, HE. Neurobiological correlates of autism: A review of recent research. *Child Neuropsychol.* 2006; 12(1): 57-79

[16] Allen, G; Courchesne, E. Differential effects of developmental cerebellar abnormality on cognitive and motor functions in the cerebellum: an fMRI study of autism. *Am. J. Psychiatry.* 2003;160(2):262-73.

[17] Friederici, AD. The neural basis of language development and its impairment. *Neuron.* 2006; 52(6): 941-952

[18] Fisher, SE; Lai, CS; Monaco, AP. Deciphering the genetic basis of speech and language disorders. *Annu. Rev. Neurosci.* 2003. 26:57-80

[19] Mulas, F; Etchepareborda, MC; Diaz-Lucero, A; Ruiz-Andres, R. Language and neurodevelopmental disorders. A review of their clinical characteristics. *Rev. Neurol.* 2006; 42(supl 2): S103-S109.

[20] Ardila, A. Las Afasias. Mexico: Universidad de Guadalajara; 2005.

[21] Talero-Gutierrez, C. Hyperlexia in Spanish-speaking children: Report of 2 cases from Colombia, South America. *J. Neurol. Sci.* 2006; 249(1); 39-45

[22] Volkmar, F; Chawarska, K; Klin, A. Autism in infancy and early Childhood. *Annu. Rev. Psychol.* 2005;56: 315-336

[23] Manning-Courtney, P; Brown, J; Molloy, CA; Reinhold, J; Murray, D; Sorensen-Burnworth, R; Messerschmitdt, T; Kent, B. Diagnosis and treatment of autism spectrum disorders. *Curr. Probl. Pediatr. Adolesc. Health Care* 2003; 33(9):283-304

[24] Talero, C; Martinez, LE; Mercado, M; Ovalle, JP; Velásquez, A; Zarruk, JG. Autismo: estado del arte. *Rev. Cienc. Salud.* 2003;1(1): 68-85.

In: Causes and Risks for Autism
Editors: A. C. Giordano et al.
ISBN: 978-1-60456-861-5

Short Communication

Histidinemia: A Risk Factor for Autism Spectrum Disorders

***Taishi Miyachi*[1], *Satoshi Sumi*[*2], *Masayuki Imaeda*[1], *Michiko Ishikawa*[1] *Hideko Morishita*[1] *and Hisako Saito*[1]**

[1] Department of Pediatrics, Nagoya City University Hospital
[2] Department of Pediatrics, West District Care Center for Disabled Children

Abstract

Histidinemia is one of the most frequent errors of amino acid metabolism. Metabolic blockage of histidase activity increases histidine concentrations in body fluids. There have been a lot of controversies whether histidinemia is a harmless biochemical error or a neurophycological disease. Several investigations demonstrated that histidinemic patients often presented autistic features, but others showed that they were asymptomatic. We investigate on 70 patients with histidinemia whether they are accompanied with autistic symptoms or not. Ten patients (14.3% of 70 patients) were diagnosed as having pervasive developmental disorder (DSM-IV). In detailed classification, 5 patients (7.1%) were autistic disorder, four (5.7%) were Asperger's disorder, one (1.4%) was pervasive developmental disorder not otherwise specified. The present study confirmed the frequency

[*] Correspondance should be addressed to : S.Sumi; West District Care Center for Disabled Children, Komoto 1-20-48, Nakagawa-ku, Nagoya 454-0828, Japan; E-mail:satoshi-sumi@mbn.nifty.com; Tel: +81-52-361-9555; Fax:+81-52-361-9560

of pervasive developmental disorder was extremely high in histidinemic patients. We hypothesize that patients should present autistic symptoms if histidinemia and other factors (whichever genetic or environmental factors) are combined; that is histidinemia may be a risk factors for autism spectrum disorders.

Introduction

Histidinemia, one of the most frequent errors of amino acid metabolism, is caused by a defect of histidase (figure 1). Metabolic blockage of histidase activity increases histidine concentrations in body fluids and decrease urocaic acid. Since Ghadimi et al. (1961) reported two cases of histidinemia with slurred and retarded speech, many cases with neurophycological symptoms have been reported. Several authors (Neville et al. 1972; Kotsopoulos et al.1979) including us (Ishikawa et al. 1987) demonstrated the relation between histidinemia and autism. Speech disturbance and school difficulty were also repoted (Lot et al.1970; Bruckman et al. 1970) On the other hand, enormous cases with histidinemia were diagnosed in the neonatal screening program in Japan (Tada et al.1982). As most of the cases showed the normal value of Intelligence Quotient (IQ) test, histidinemia was regarded as a harmless disorder as far as mental retardation is concerned.

Autism is a behaviorally defined syndrome, characterized by pervasive impairments in social interaction and communication, and the presence of stereotype. In this decade the diagnostic criteria of autism has been expanded, from a strict category (Kanner type) to a broad spectrum, owing the progress of neurophycological understanding (Wing L, 1996). The broad spectrum of autism is defined as pervasive developmental disorder (PDD) in DSM-IV criteria (American Psychiatric Association 1994) and the number of cases has recently increased rapidly, in line with the expanding criteria. A recent study described the prevalence in the general population to be 1 to 2% (Baird et al.2006; Posserud et al. 2006; Sumi et al. 2006). However, the pathological mechanism of autism has not been clarified yet. Until now it has been proved that a small number of whole autistic patients are associated with known diseases. For example, about 20% of untreated patients with phenylketonuria (PKU), a disorder of amino acid metabolism, are also diagnosed as having autism (Reiss et al. 1986). A lot of combinations of multi-factors maybe cause the same behavioral dysfunction by the impairment in common nerve pathway.

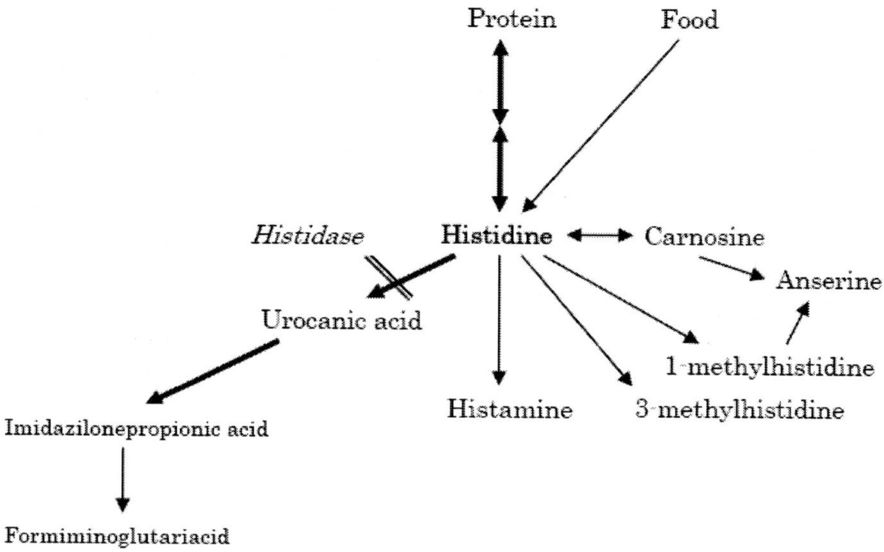

Figure 1. Metabolic pathway of histidine.

In this report, we invest clinical and biochemical observations of 70 cases with histidinemia to clarify the relation between this disorder and autism.

Patients and Methods

1. Biochemical Diagnosis

In the neonatal screening program, patients having high histidine levels (more than 6 mg/dl, measured by the Guthrie method) were examined at Nagoya City University Hospital. Amino acid including histidine were measured at SRL laboratory (Hachiohji, Tokyo) using the amino acid analyzer. The diagnosis of histidinemia was made by the repeated findings of a fasting plasma concentration of histidine above 8 mg/dl. Among 109 patients who were detected in the neonatal program, we diagnosed 106 patients as having continual histidinemia .All patients were Japanese.

2. Neurodevelopmental Evaluation and Diagnosis of Pervasive Developmental Disorder

After careful observations over 5 years after birth, we diagnosed patients. Detailed patient's histories were obtained from parents, with paying attention to the autistic symptoms. Each patient was interviewed at least 5 times. Developmental quotient (DQ) (below 3 years old) or intelligence quotient (IQ) (over 3 years old) was assessed by Tsumori-Inage method or by Wechsler Intelligence Scale for Children Revised (WISC-R). Among106 patients, we could observe 70 patients according to the above schedule. As we could not continue to follow up other cases, we excluded them in this study. We excluded the 27 patients in this study. We employed DSM-IV (American Psychiatric Association, 1994) criteria for diagnosing pervasive developmental disorder and other neurodevelopmental disorders. Finally they were classified to autistic disorder (AS), Rett's disorder, childhood disintegrative disorder, Asperger's disorder (Asp) and pervasive developmental disorder not otherwise specified (PDD-NOS), learning disorder (LD), attention deficit/hyperactivity disorder (ADHD), borderline IQ and normal.

Table 1. Frequency of pervasive developmental disorder in 70 patients with histidinemia

Diagnosis	Number	Frequency	IQ(mean±S.D.)	histidine(mg/dl)
AS	5	7.1%	89.8±31.2	13.0±1.6
Asp	4	5.7%	104.2±7.29	12.4±1.09
PDD-NOS	1	1.4%	107	12
LD	4	5.7%	88.3±7.8	12.6±3.0
ADHD	2	2.8%	95.5±2.5	14.1±3.12
Borderline	5	7.1%	79.4±4.2	12.4±2.0
Normal	49	70%	106.7±12.7	12.7±2.6
Total	70	100%	101.8±16.4	12.6±2.6

PDD: pervasive developmental disorder, AS: autistic disorder, Asp: Asperger's disorder, PDD-NOS: pervasive developmental disorder not otherwise specified, LD: learning disorder, ADHD: attention deficit/hyperactivity disorder, borderline: borderline IQ.

Results

As shown in table 1, 10 patients (14.3% of 70 patients) were diagnosed as having pervasive developmental disorder, according to the criteria of DSM-IV. In a detailed classification, 5 patients (7.1%) were AS, four (5.7%) were Asp, one (1.4%) was PDD-NOS. Two patients with AS have already reported in our previous report (Cases III and IV in Ishikawa et al, 1987).

Conclusion

There have been a lot of controversies whether histidinemia is a harmless biochemical error or disease. Neville et al reported 2 of 7 histidinemic patients showed characteristic features of autism (Neville et al.1972). Ghadimi et al (1961) also described behavior disturbances such as “flighty attitude”, “bad temper”, scholastic failure, speech impairment, and mild degree of retardation. Kotsopoulos et al (1979) and we also reported several histidinemic patients with autistic features (Kotsopoulos et al. 1979). However, the studies of the neonatal screening program (Tada et al. 1982; Scriver et al. 1983) have emphasized the normality of IQ in histidinemia.

On the other hand, the concept of autistic disorder has drastically changed in this decade. Autism used to be regarded as an extremely rare disorder manifested by severe impairments in social interaction and communication. Most of these patients were accompanied with mentally retardation. But the criteria of autism has been expanded, from a strict category to a broad spectrum disorder including normal IQ patients. We believe it is necessary to invest histidinemia again in a view point of the present concept of autism.

In our patients the result of IQ test was 101.8±16.4 (mean±S.D.). Only one patients showed low IQ score (below 80), corresponding with the previous reports. Namely, histidinemia seems to be harmless as far as mental retardation. However, we demonstrated that about 14% of histidinemic patients presented autistic features, corresponding to the criteria of pervasive developmental disorder. The frequency of pervasive developmental disorder in histidimemia is much higher than that in normal populations. But we also confirmed that a lot of histidinemic patients can develop normally. The reason of this wide clinical heterogeneity is still unknown. We could not find any relationship between histidine concentrations and clinical symptoms. We also carried out a molecular analysis on histidinemic patients in order to examine a relationship between mutations in histidase gene and clinical symptoms. However, we have not able to find a clear relationship yet (Kawai et al., 2006). Therefore, we hypothesize that patients should present autistic symptoms if histidinemia and other factor (whichever genetic or environmental factor) are combined. Namely histidinemia may be a risk factor for autism spectrum disorders. Further investigations are necessary to clarify the mechanism of autistic symptoms in histidinemia.

Acknowledgement

The authors would like to thank the late Prof Yoshiro Wada, Drs Toshiyuki Iguchi and Tomoko Asai, Psychologists Hisayo Imahashi, Rie Yamada, Masae Endo and Akiko Goto for their supoort in this study.

References

American Psychiatric Association. (1994) Diagnostic and statistical manual of mental disorders:DSM-IV. 4th edn. Washington, American Psychiatric Association.

Baird G, Simonoff E, Pickles A, Chandler S, Loucas T, Meldrum D, Charman T(2006) Prevalence of disorder of the autism spectrum in a population cohort of children in South Thames: the special needs and project. *Lancet.* 368, 210-215.

Bruckman, C., Berry, H.K., Dasenbrock, R. (1970) Histidinemia in two successive generations. *Am. J. Dis. Child.*119, 221-227.

Ghadimi, H., Partington, M.W., Hunter, A. (1961) A familial disturbance of histidine metabolism. *N. Eng. J. Med.* 265, 221 only.

Ishikawa, M.,Saito, H., Morishita, H., Ito, T., Kato, T., Wada, Y. (1987) Development of histidinemic patients:follow-up study of five cases with neuropsychological symptoms. *Acta Paediatr. Jpn.* 29, 444-448.

Kawai, Y., Moriyama, A., Asai, K., Coleman-Campbell, C.M., Sumi, S., Morisihita, H., & Suchi, M. (2005) Molecular characterization of histidinemia:identification of four missence mutations in the histidase gene. *Hum. Genet.* 116, 340-346.

Kotsopoulos, S., Kutty, K.M. (1979) Histidinemia and infantile autism. *J. Autism.*

Dev. Disord. 9, 55-60.

Lot, I.T., Wheelden, J.A., Levy, H.L. (1970) Speech and histidinemia: methodology and evaluation of four cases. *Dev. Med. Child Neurol.*12, 596-603.

Neville, B.G.R., Bentonim, A., Clayton, B.E., Shepherd, J. (1972) Histidinemia;study of relation between clinical and biological findings in 7 subjects. *Arch. Dis. Child.* 47, 190-200.

Posserud MB, Lundervold AJ, Gillberg C (2006) Autistic features in a total population of 7-9-year-old children assessed by ASSQ(Autism Spectrum Screening Questionnaire) *J. Child Psychiatry.* 47: 167-75.

Reiss, A.L., Feinstein, C., Rosenbaum, K.N. (1986) Autism in genetic disorders. Shizophrenia Bulletin, 13, 724-738.

Scriver, C.R., Levy, H.L.. (1983) Histidinemia. Part 1: reconciling retrospective and prospective findings. *J. Inher. Metab. Dis.* 6, 54-57.

Tada, K., Tateda, H., Arashima, S., Sakai, K., Kitagawa, T., Aoki, K., Suwa, S., Kawamura, M., Oura, T.,Takesada, M.,Kuroda, Y.,Yamashita, F., Matsuda, I., Naruse, H. (1982). Intellectual development in patients with untreated histidinemia. A collaborative study group of neonatal screening for inborn errors of metabolism in Japan. *J. Pediatrics.* 101, 562-563.

Sumi S, Taniai H, Miyachi T, Tanemura M.(2006) Sibling risk of pervasive developmental disorder estimated by an epidemiologic survey in Nagoya, Japan. *J. Hum. Genet.* 51:518-522.

Wing, L. (1996) The autism spectrum. A guide for parents and professionals. London, Constable and Company Ltd.

Index

D

H

I

K

L

M

N

O

P

Q

R

S

T

U

V

W

X

Y